INTRODUCTION TO
Computer Mathematics

COMPUTERS AND MATH SERIES
Marvin Marcus, Editor
University of California at Santa Barbara

L. A. Carmony et al.
Problem Solving in Apple* Pascal

L. A. Carmony et al.
Problem Solving in Apple* Pascal, Student's Diskette

L. A. Carmony et al.
Problem Solving in Apple* Pascal, Teacher's Guide and Solution Manual

M. Marcus
Discrete Mathematics: A Computational Approach Using BASIC

M. Marcus
Discrete Mathematics Diskette

R. Merris
Introduction to Computer Mathematics

S. G. Williamson
Combinatorics for Computer Science

*Apple is a trademark of Apple Computers, Inc.

INTRODUCTION TO
Computer Mathematics

RUSSELL MERRIS

*Department of Mathematics
and Computer Science
California State University, Hayward*

COMPUTER SCIENCE PRESS

Computer Science Press
1803 Research Boulevard
Rockville, Maryland 20850
1 2 3 4 5 6 Printing   Year 89 88 87 86 85

Library of Congress Cataloging in Publication Data

Merris, Russell, 1943–
    Introduction to computer mathematics.

    Includes index.
    1. Mathematics—Data processing. 2. Basic (Computer program language) I. Title.
QA76.95.M47 1985   510'.28'54   85-4683
ISBN 0-88175-083-2 (Student Textbook)

CONTENTS

PREFACE

This book provides a thorough introduction to computer programming in the BASIC language. Yet the computer, in and of itself, is not the primary object of study. The central theme is the relationship between computers and mathematics. These two topics are presented in an integrated way. The computer is exploited both as a motivational device and as a powerful tool in the study of probability, statistics, algebra, and geometry. Programming skills are developed naturally as the need for them arises.

Of course, the computer is not equally useful in all areas of mathematics. The machine has, in a way, dictated the selection of topics for this book. In the same manner, computers have produced new constraints and imperatives on the curriculum itself. It seems important for those with a mathematical inclination to take a hand in reshaping the course of study so that computers will be incorporated in the most appropriate way.

Hayward, California
January 1985

Dedicated to
Alice Gill and Michael Merris

Chapter 1
INTERACTIVE COMPUTATION

1.1 IN THE BEGINNING

Once the machine is "up" and running, the screen may contain various messages. Apart from these, you will observe a bright spot. (It may be a square, a rectangle, or a short line segment. It may be flashing on and off, or it may be continuously on.) This is called the cursor (CRSR). It indicates where on the screen the computer's attention is focused. The screen is the means by which the computer will communicate with you.

You communicate with the computer by means of the keyboard. It is very much like a typewriter. Begin by using it to type your first name. If you make a mistake, look for a backspace key or a DELete key. As you type along, notice how the cursor moves. Now type a space followed by your last name. Once you've gotten your name written, press the RETURN (or ENTER or ←) key. Below your name will appear a message, perhaps SYNTAX ERROR. This does not mean the computer has rejected your name. It simply means that the machine is confused. This may surprise you, but it is certainly true. The problem is that computers have a very limited vocabulary. At the moment, it doesn't have a clue as to what is going on. In such situations, it has been programmed to respond with an "error" message.

The most important thing to know about computers is that they have no imagination at all. Within the limits of its ability, the computer will do what you tell it to do—*exactly* what you tell it to do. Our literature contains many stories dealing with just this stubborn tendency to interpret instructions literally. You are probably familiar with "King Midas and the Golden Touch," or "The Three Wishes." The computer is a faithful servant which will try to execute, both literally and exactly, those of your instructions that it understands. It will have neither the intelligence nor the imagination to discern what you really meant it to do.†

Fortunately, computers do not respond to instructions until they are confirmed. A command is confirmed by pressing the RETURN, ENTER, ←, or similar key on your keyboard. *In the remainder of the book, the word* ENTER *will be used to refer to this key, even though that may not be the correct word for your machine.* The computer takes "official" note of what you have typed only after you press the ENTER key. This feature gives you (by means of backspacing or DELeting) the opportunity to correct any inadvertent typing errors before giving a formal instruction. (Remember King Midas!)

Let's have pity on the machine and give it some instructions it can understand. One of the things computers do best, of course, is compute. Try typing the following line:

    3 + 5

What we want from the computer is the sum of 3 and 5. Confirm the command by pressing ENTER.

Don't be surprised that nothing happened or, more accurately, that nothing appeared to happen. One (illustrative though not entirely accurate) way to imagine the computer's response is this: Adding 3 to 5 it gets 8, thus fulfilling our instruction to the letter. We didn't instruct it to share the answer with us and it didn't have the imagination to do so on its own!

Try typing this line followed by ENTER.

    PRINT 3 + 5

Now we're getting somewhere. PRINT is one of the words in your computer's limited vocabulary. PRINT may be thought of as a code word. The computer interprets the PRINT command as an instruction to make what follows it appear on the screen. Items ENTERed from

---

†One beginning student was heard to exclaim, "Computers never do what you want them to do, only what you tell them to do."

the keyboard are sometimes referred to as input. The computer's responses are called output. PRINT is an instruction to output the result of the calculation.

Try inputting this:

```
PRINT  "3+5"
```

(Don't forget to confirm your instruction when you are satisfied that you have typed it in correctly.) As you see, characters enclosed in quotes are PRINTed exactly as typed. Now try this (ENTER it exactly as shown).

```
PRINT  "3+5  ="  3+5
```

Machines differ as to punctuation. If you ENTERed the instruction exactly as shown and got the SYNTAX ERROR message, try adding a semicolon:

```
PRINT  "3+5  =" ; 3+5
```

Try some other arithmetic computations, for example,

```
PRINT  5+9
PRINT  9—2
```

(A hyphen on the typewriter keyboard is interpreted as a "minus sign" by the computer. Multiplication is rendered by the symbol $*$, while division is accomplished by means of the slash: /.)

```
PRINT  3*5
PRINT  3/5
```

See what happens when you punctuate using semicolons and commas as in

```
PRINT  3+5 ; 3—5 ; 3*5 ; 3/5
```

or

```
PRINT  60/10 , 60/12 , 60/15
```

Make up some examples of your own, but resist the temptation to just have the computer multiply and divide big numbers. It's interesting to see what happens, and the computer will probably manage just fine, but you won't learn anything. At present, the important thing is for you to learn how to communicate with the computer and, perhaps just as important, for you to gain some confidence that you are actually communicating what you think you are communicating. Concentrate on asking the machine to do only those problems that you don't mind checking on your own with paper and pencil.†

Use any remaining time you have to become familiar with some of the special keys on your keyboard. Make a mental note of the response, if any, accorded to each of these keys.

Exercises (1.1)

1. Have the computer perform the indicated operation and confirm the result using paper and pencil.
   a. 21*37   b. 15*823   c. 231*481   d. 273*407

2. There are 3.785 liters in a gallon. If you put 53 liters of gasoline into your car, how many gallons have you put in? (Use the computer.)

3. There are 2.54 cm (centimeters) in an inch. PRINT the number of centimeters in
   a. 6 inches   b. 1 foot   c. 1 yard   d. ½ inch

4. Assume that it costs 20 cents a mile to operate your car. How much does it cost to drive 25 miles? (Check the computer with paper and pencil.)

5. In California, the State sales tax is 6%. If you buy a stereo amplifier for $169.95, what will the total bill come to? (*Hint:* If you multiply by .06, you will compute the tax, which must then be added to $169.95 to obtain the answer. What happens if you multiply by 1.06?)‡

6. In certain counties of California adjacent to the San Francisco Bay, the sales tax is 6.5%. (The extra 0.5% is used to help finance BART, the Bay Area Rapid Transit system.) If you buy the same $169.95 amplifier in one of these counties, what will the total bill come to?

---

†Do not erase anything in the vicinity of the computer. Dust particles of any kind are bad for the internal machinery.

‡ If you get an error message, it may be that you are trying to ENTER the percent and/or dollar signs. These will confuse the computer. (You must convert the percent to a decimal on your own.) Other common errors are to type lower case "el" instead of "one," or "oh" for "zero."

7. (Refer to Exercises 4–6 while doing this problem.) Assume you can avoid the extra 0.5% tax by making your purchase in the next county. If this involves a 25-mile (round) trip, how much money will it save you? Try two different solutions to this exercise. The first is to use paper and pencil with the answers to Exercises 4–6. The second is to try the following:

```
PRINT 169.95*.005 -.20*25
```

Explain why the answers should be the same. If they aren't, try to find your error(s).

8. What retail price would justify a trip to the next county to avoid the BART tax? (See Exercises 4–7.)

9. In California, the sales tax is assessed on the purchase of new automobiles. How much would you save by purchasing a $9950.00 car in a county in which the BART tax is not assessed? (See Exercises 4–7.)†

1.2 MEMORY AND PRECEDENCE

One hears that, apart from computing, one of the things computers do best is remember. The computer has a vast memory. You might think of it as thousands of open, empty egg cartons. Each space in each egg carton is a memory location. The computer "remembers" a piece of information by putting it into one of these memory locations—a random one. You may have heard the term RAM which is "computerese" for Random Access Memory. (Anything stored in RAM is lost when the computer is turned off.) What all this means for us is that we not only have to tell the computer to remember something, we even have to tell it how to recall that something from its memory! If this sounds complicated, it isn't. In practice, we just tell the computer to label the location in its memory where it has stored each piece of information. For example, if we want our computer to remember the product of 3 and 5, we simply instruct it to label the answer. ENTER this instruction:

```
LET A=3*5
```

---

†When a number such as 9,950.00 is ENTERed, the comma must be omitted.

"LET" is another computer code word. The computer interprets our instruction as follows: Multiply 3 and 5, store the answer in some memory location, and remember where you put it by giving the location the label "A." You may test the machine's memory by ENTERing this instruction.

    PRINT A

We can use the symbol A in algebraic expressions as if it were 15. For example, try PRINTing A − 5 or 2*A. Suppose we

    LET B = 4

and then

    PRINT A*B

On most computers, the LET is optional. That is to say, one may simply ENTER

    C = 3 + 4

While this looks like an algebraic equation to you, it is interpreted by the computer exactly as if you had ENTERed

    LET C = 3 + 4

There is an honest difference of opinion among programmers about the value of using LET. One camp feels it makes the idea of labeling memory locations easier to understand and makes computer programs easier to read. Others are of the opinion that since LET is not necessary, it is a superfluous decoration. In this book, LET will not be used, but you should feel free to use it yourself. See what happens when you ENTER this sequence of instructions.

    A1 = 3
    A2 = 4
    A3 = A1 + A2
    PRINT A3

If you wish to change the contents of a memory location, you may do so. The following sequence of instructions will illustrate the technique.

    X = 3
    PRINT X
    X = 4
    PRINT X

You might ask how X can equal 4 when it already equals 3. If 3 equals X and X equals 4, shouldn't 3 equal 4? The answer lies in the interpretation given to the equal sign by the computer; X is simply a label for a memory location. To the computer, X does not equal 3 any more than your house is equal to your street address. Of course, what the computer has put into location X, it can change. If you are having trouble with this idea, it might help to rewrite the last sequence of instructions using the optional LET.

Every politician who is up for reelection is fond of the slogan, "Don't change horses in the middle of the stream." Let's hope that idea does not extend to changing topics in the middle of a chapter because that's what we are about to do. If you're not sure you fully understand the discussion of memory, reread it before going on.

Type the following instruction, but don't ENTER it yet.

```
PRINT 2*3+4*5
```

Let's try to anticipate the computer's response, remembering that it will try to do exactly what it has been told to do. It seems we should multiply 2 by 3 to get 6. Then add 4 to obtain 10. Finally multiply by 5 to reach the answer, namely, 50. This, of course, is not what was intended. In an algebraic expression of the form 2*3 + 4*5, it is our convention to do the multiplications first and then the addition. This convention produces the result 6 + 20 = 26. Press the ENTER key, and see which result the computer comes up with.

Contrary to our expectation, the computer did what we wanted it to do and not what we told it to do. Why? The answer is that your computer has previously been told about (some of) our conventions for doing a sequence of arithmetic operations. In addition to RAM, the computer has another area of memory called ROM: Read Only Memory. ROM is permanent memory installed at the factory. Unlike RAM, ROM is not lost when the computer is turned off. It is remotely like the part of your brain that controls the involuntary bodily functions (such as heartbeat).

Given a sequence of arithmetic operations to perform (i.e., the operations +, −, *, and /), the computer will first do the multiplications and divisions as it comes to them, reading from left to right. It will then make a second pass along the line to do the additions and subtractions, also in the order in which it encounters them while reading from left to right. Using these rules of precedence, predict the response to the following line before you ENTER it:

PRINT 1 / 2 * 5

You have the power to override the usual precedence by using parentheses. ENTER this line and see what happens.

PRINT 1 / (2 * 5)

The parentheses tell the computer to treat 2*5 as a single number, in effect telling it to do the multiplication before the division. Type in each of the following lines and try to predict the computer's response before ENTERing.

PRINT 2 * 3 / 6 * 5
PRINT (2 * 3) / (6 * 5)
PRINT 2 * (3 / 6) * 5
PRINT 2 * 3 / (6 * 5)
PRINT 12 / 2 / 3
PRINT 12 / (2 / 3)

Think up a few problems of your own, work out the answers using paper and pencil, then ENTER them to see how the computer responds. But, keep your problems simple. The danger is that if the problem is too difficult, you will make a mistake. In that case, it is most unlikely that you and the computer will agree. If there is an unexpected discrepancy, check your paper and pencil work carefully.

Exercises (1.2)

1. What responses do you *expect* the computer to make to these instructions?
   a. PRINT 1 / 2 / 3 * 4 / 5 * 6
   b. PRINT 1 + 2 * 3 / 4 * 5 / 6
   c. PRINT 1 / 1 + 2 / 2 + 3 / 3 + 4
   d. PRINT (1 + 2) * 3 / (4 * 5) * 6

2. What response does the computer make to each of the instructions in Exercise 1?

3. Given that there are 365.25 days in a year, write a computer instruction to determine how many years there are in 1 million days.

4. ENTER your instruction from Exercise 3, and use the answer to determine what year it was 1 million days ago.

5. An "astronomical unit" (AU) is the average distance from the Earth to the Sun. One AU is 149,500,000 km (kilometers). Write a computer instruction to determine the number of miles in an AU. (*Hint:* There are approximately 1.6 km in a mile.)

6. One horsepower is about 745 watts. If the power from a 75 horse-power outboard motor were converted (without loss) to electrical energy, how many 100-watt bulbs could it operate at full capacity. (Write down a computer instruction for finding the answer, EN-TER it, and record the response.)

7. To convert from the Fahrenheit temperature scale to the centi-grade (or Celsius) scale, subtract 32 and multiply the result by 5/9. Write computer instructions for converting the following Fahrenheit temperature readings to centigrade readings:
   a. 70   b. 100   c. 32   d. 212
   (At sea level, water freezes at 32 and boils at 212 degrees F.)

8. ENTER each of your instructions from Exercise 7, and record the answers.

9. Describe in words how to convert from the centigrade scale to Fahrenheit. (*Hint:* Reverse the conversion described in Exercise 7.)

10. Write a computer instruction to convert 35 degrees centigrade to Fahrenheit.

11. ENTER your instruction from Exercise 10, and record the answer.

12. A stack of 100 brand new $1 bills is approximately 1-cm (centi-meter) thick. Write a computer instruction to determine how many miles thick a stack of 1 billion new $1 bills would be. (*Hint:* There are 2.54 cm in an inch and 5280 feet in a mile. Assume the bills at the bottom are not compressed by the weight.)

13. ENTER your instruction from Exercise 12, and record the answer.

Suppose you have lunch with an exchange student from France. In the course of the meal, the conversation turns to the subject of fuel economy. When you tell her that your car gets 22 miles per gallon on the highway, she has to stop and think for a minute. Used to dealing with kilometers and liters, it's a little difficult for her to grasp whether

22 m.p.g. is good or bad. Let's use the computer to help her by converting our units to hers.

Unfortunately, the computer is even worse off than the girl from France. Not only is it not familiar with m.p.g., it has never heard of liters or kilometers ... or even of cars! It turns out that there are approximately 1.6 km in a mile and 3.785 liters in a gallon. The computer will happily do the arithmetic, but *we* have to set up the problem. A very useful trick in this context is to treat the units as if they were fractions. Treat "miles per gallon" as the fraction miles/gallons and "kilometers in a mile" as the fraction kilometers/miles. Now comes the fancy part. What do you suppose happens if you "multiply"

(kilometers/miles)*(miles/gallons)?

The answer is that the "miles" cancel, resulting in kilometers/gallons, i.e., km per gallon! Thus, if we multiply 1.6*22, we obtain the "mileage" in units of km per gallon. If we could just arrange to multiply the result by gallons/liter, we would be finished. Now, "liters in a gallon" converts to the fraction liters/gallon, not gallons/liter. Thus, the appropriate thing to do is not multiply but divide. If we divide by liters/gallon, we invert and multiply by gallons/liter. Thus, we want to

```
PRINT 1.6*22/3.785
```

(The answer cannot be more accurate than the least accurate number involved in the computation. In this case, both 22 and 1.6 are accurate only to two digits. Thus, in spite of the many decimal places produced by the computer, the only meaningful answer is 9.3 km per liter.)

14. Write a computer instruction to convert 35 miles per gallon to kilometers per liter. (*Hint:* See the discussion following Exercise 13.)

15. Enter the instruction from Exercise 14 and record
    a. the computer's response
    b. the most appropriate answer

16. Write a computer instruction to convert 15 km per liter to miles per gallon.

17. ENTER the instruction from Exercise 16 and record
    a. the computer's response
    b. the most appropriate answer

1.3 EXPONENTIATION

How many ancestors do you have 20 generations back? One generation back, you have 2, your (biological) parents. Since each of your parents had 2 parents, your grandparents, you have a total of $2 \times 2 = 4$ ancestors 2 generations back. And, since each of these ancestors had 2 parents (your great-grandparents), you have a total of $2 \times 4 = (2 \times 2 \times 2$ $=)$ 8 ancestors 3 generations back.

A pattern is beginning to emerge. Four generations back, you had $2 \times 2 \times 2 \times 2$ ancestors, and so on. Finally, you had 2-times-itself-20-times ancestors 20 generations back. The following shorthand notation is used to indicate this quantity:

$$2^{20}$$

This expression is read "2 to the 20th power," or just "2 to the 20th." The 20 in the expression is called an *exponent*, and the process by which 2 is "raised" to the 20th, or any other power, is called exponentiation.

Your computer has an exponentiation key. On some computers, the exponentiation symbol is a small dunce cap. On others, it is an arrow pointed to the top of the keyboard.[†] Look for the key on your computer. When you have found it, ENTER this:

```
PRINT 2↑20
```

Does it surprise you to find there were so many? If we assume 4 generations in a century, then it seems there were more than a million of your ancestors around when Columbus discovered America! The surprise is due in part to the fact that the sequence 2, 4, 8, . . . , starts so innocently. It's hard to believe the 20th term is already more than 1 million. Exponentiation is a way to produce big numbers in a hurry. (Surely, there is no danger of confusing 2^{20} with 40.) Experiment further by ENTERing these lines

---

[†]On the Apple II, SHIFT N is the exponentiation key. On the IBM PC and the Apple 2c, it is SHIFT 6. On the Commodore 64, it is the "up arrow" (next to the "RESTORE" key). In this book, the symbol ↑ will be used.

```
PRINT 2↑10
PRINT 10↑3
PRINT 10↑6
PRINT 10↑20
```

The last line produces a mysterious response. What does "E + 20" mean? Evidently, it is the computer's way of expressing $10 \uparrow 20$, that is, 10^{20}. The expression "E + 20" means "multiplied by $10 \uparrow 20$," or "move the decimal point 20 places to the right." In responding to PRINT $10 \uparrow 20$, the computer paraphrased the question rather than answering it. We'll have more to say about this in a minute.

It is a useful rule of thumb to remember that $2 \uparrow 10$ is approximately $10 \uparrow 3$, i.e., 1024 is not far from 1000. (Which is bigger, 10^2 or 2^{10}?) This approximation can help us, for example, with the following problem: What is the relationship between 2^{10} and 2^{20}? The first is the product of 2 times itself 10 times, while the second is the product of 2 times itself 20 times. Indeed, $2^{20} = (2^{10})^2$, in other words, 2^{10} times itself. Thus, 2^{20} is approximately 1000 times 1000, that is, 1 million ... exactly what we have seen. By the same token, 2^{30} is $(2^{10})^3 = (2^{10})*(2^{10})*(2^{10}) = (2^{10})*(2^{20})$. In particular, 2^{30} is approximately 1000 times 2^{20}, or 1000 million. [†] Enter this instruction

```
PRINT 2↑30
```

Once again, we encounter the mystery response, but this time we're ready for it. We interpret 1.07374182E + 09 as 1,073,741,820, i.e., 1 billion, 73 million, 741 thousand, 820. (The last digit, 0, is suspect. The computer reports only to 8 decimal places.)

While you may know that a million million is a trillion, what is the name for a trillion trillion? If you don't know, does that make you dumb? Of course not! There are a great many very smart, very well-educated people who couldn't name a trillion trillion, even though they might routinely deal with numbers that size or much bigger. People generally use "scientific notation" to deal with really large numbers. Scientific notation is just a fancy way to talk about exponentiation.

[†] In American usage, 1 thousand million is 1 billion. In British usage, 1 thousand million is just that, a thousand million. The British "billion" is 1 million million, what Americans call 1 trillion.

Instead of saying there are 149,500,000 km in an astronomical unit, astronomers would say there are $1.495*10^8$ km in an astronomical unit. Indeed, a scientist would generally avoid saying even "a trillion," preferring "10 to the 12th" instead. This, of course, is just what your computer is trying to imitate with it's somewhat pitiful "E + 09" and "E + 20".

Let's explore what precedence exponentiation has. Try ENTERing these instructions:

```
PRINT  3*4↑2
PRINT  3↑2*4
PRINT  −2↑4
PRINT  2↑7/2↑4
```

It turns out that exponentiation takes precedence over all four arithmetic operations. Look at the response to PRINT 3*4 ↑ 2. There are two possibilities. One is $(3*4)^2$, i.e., 12^2,† and the other is $3*(4^2)$. The computer did the second of these. It first computed 4 squared, got 16, and then multiplied by 3 to obtain 48. Notice that it did the exponentiation first, even though the first operation it came to was the multiplication. Responding to PRINT 3 ↑ 2*4, the computer evaluated 3^2, got 9, and multiplied that by 4. The response to the third instruction will depend on your computer. Some, e.g., the IBM PC and the Commodore 64, interpret the " − " as indicating subtraction and compute $-(2^4)$. Others, e.g., the Apple II, interpret the " − " as meaning "negative 2" and compute $(-2)^4$. Finally, what about $2^7/2^4$? The computer is going to do the exponentiations first and then the division: 128/16. But we don't need a computer for this one! If we have a product of seven 2's divided by four 2's multiplied together, why not just cancel four 2's from the numerator and denominator? We're left with $2^3 = 8$.

One last convention requires our attention. ENTER these lines.

```
PRINT  1/100
PRINT  10↑(−2)
PRINT  1/10↑20
PRINT  1/2↑20
```

---

†We usually read 12^2 as "12 squared" because that is the computation one does to find the area of a square 12 units on a side. Similarly, we'll read 4^2 as "4 squared" rather than "4 to the 2nd." Analogously, 4^3 is read "4 cubed."

The expression "E − 07" in the response "9.53674317E − 07" should be interpreted as "move the decimal point 7 places to the LEFT" or "*divide by* 10^7." It is convenient to define

$$10^{-7} = 1/10^7$$

Similarly,

$$2^{-5} = (1/2)^5 = 1/(2^5)$$

The negative sign in the exponent indicates reciprocation.

Microcomputers generally "think" in terms of 8 decimal places. Suppose instead of responding to PRINT $1/2\uparrow 20$ with 9.53674317E − 07, the machine had output 0.00000095. Would this make us happier? It would certainly have a more familiar appearance. But, the price would be throwing away 7 digits of accuracy. While it may be that only two digits are meaningful to us, we want to retain the option of making that decision for ourselves. Thus, we have to be prepared to deal with negative exponents.

Exercises   (1.3)

1. Which is bigger, 2^3 or 3^2? (This is a problem for you, not the computer!)

2. Enter the following instructions:

   ```
   PRINT 3↑2−2↑2
   PRINT 4↑2−3↑2
   PRINT 5↑2−4↑2
   PRINT 6↑2−5↑2
   ```

   Do you see a pattern emerging? Formulate a conjecture about the difference of successive "squares."

3. Predict the response if the computer were instructed to PRINT
   a. 1+3↑3∗4/6
   b. 8/2↑4∗6−2
   c. 10↑3↑2
   d. 3↑2+4↑2−5↑2
   e. (3+4−5)↑2
   f. (3↑2+4↑2)/5↑2
   g. −2↑3
   h. −3↑2
   i. (5↑2+12↑2)/13↑2

4. Write down an estimate for 2^{23}, for 2^{32}, for 2^{41}. (Don't do any actual

arithmetic, and don't use the computer. Base your estimates on the fact that 2^{10} is approximately 10^3.)

5. Without doing any adding, make a guess for the sum

$$1 + 3 + 5 + 7 + \cdots + 33$$

(*Hint:* The sum of the first 17 odd numbers is a "perfect square." Use your solution to Exercise 2 to guess which one.)

6. Compute each of the following (on your own, without the computer).
   a. 4^4 b. 5^3 c. 2^7 d. 3^3 e. $(-3)^3$
   f. $(1/2)^3$ g. 1^{100} h. 2^8 i. 2^9 j. 2^{-6}
   k. $(1/3)^2$ l. $(3/5)^2$

7. A number in scientific notation has the form: a single digit, followed by a decimal point, followed by more digits, times some power of 10. These numbers are all written in scientific notation:
   1 3.14 $1*10^{10}$ $6.02*10^{23}$ $6.625*10^{-27}$ $-1.23*10^5$
   These numbers are not written in scientific notation:
   $5{,}280$ $1{,}495*10$ $22/7$ $1/10$ $1/2$
   Write each of these last five numbers in scientific notation.

8. Convert each of the following from scientific notation to ordinary notation:
   a. $7.1*10^8$ b. $6.02*10^{23}$ c. $6.625*10^{-27}$

9. Interpret each of the following computer responses as ordinary numbers (without using exponents).
   a. $1.234E+07$ b. $1.234E-05$ c. $1.234E-03$
   d. $1.234E+11$

10. Tabulate the first 20 powers of 2. (*Hint:* the first is $2^1 = 2$, the last is $2^{20} = 1{,}048{,}576$.) Save these numbers for later use.

11. On your own, without the computer, express each of the following either as a power of 2 or as the reciprocal of a power of 2.
    a. $2^{17}/2^{12}$ b. $2^{13}/2^6$ c. $2^{217}/2^{109}$ d. $2^{63}/2^{80}$

12. The *digital root* of a number is obtained as follows: Sum the digits comprising the number. If the sum is a single digit, then it is the digital root. Otherwise, sum its digits, obtaining a still smaller number. If this answer is a single digit, it is the digital root of the original number. If not, sum its digits. Continue this process until

a single digit is reached. For example, to compute the digital root of 78,964, we first compute $7 + 8 + 9 + 6 + 4 = 34$ followed by $3 + 4 = 7$. Thus, the digital root of 78,964 is 7. Did you know that a number is evenly divisible by 3 if and only if its digital root is divisible by 3? Using this fact, determine which of the following numbers is divisible by 3.

a. 15,872   b. 167,593,221   c. 693,555,721

13. Tabulate the squares of the first 30 numbers. (Double check your work.) Compute the digital root of each of these "perfect" squares. (See Exercise 12. For example, the square of 27 is $27^2 = 729$. The digital root of 729 is 9.) See if you can detect a pattern. Use it to decide which numbers can be the digital root of a perfect square. List them. Finally, see if you can determine whether 123,456,789,050,000 is a perfect square.

14. The speed of light in a vacuum is approximately 186,000 miles per second. How far can a quantum of light travel in a year? (Even the astronomical unit is too small to be useful in discussions of interstellar distances. Astronomers have generally adopted the light-year, namely the distance light travels in a year, as a more convenient unit. With respect to this unit, Alpha Centauri, the *closest* star to our Sun, is thought to be 4.3 light-years away!)

15. The surface area of a sphere of radius r is $4\pi r^2$, where π is about 3.14159. Find the surface area of a sphere of radius
a. 2 feet   b. 3 meters   c. 4 yards   d. 4.3 light-years

16. The volume of a sphere of radius r is $(4/3)\pi r^3$. Find the volume of a sphere of radius
a. 2 feet   b. 3 meters   c. 4 yards   d. 4.3 light-years

17. The area of the curved surface of a (right circular) cylinder is $2\pi rh$, where r is the radius of the base and h is the height. Find the curved surface area if
a. $r = 2$ inches and $h = 16$ inches
b. $r = 16$ astronomical units (AU) and $h = 2$ AU
c. $r = 4$ cm and $h = 4$ cm
d. $r = 1.5$ meters and $h = 15$ cm
(A centimeter, abbreviated cm, is 0.01 meter.)

18. The total surface area of a cyclinder is the area of the curved surface plus the areas of the circular top and bottom. Compute the total surface area for each part of Exercise 17.

19. The volume of a (right circular) cylinder is $\pi r^2 h$. Compute the volume for each part of Exercise 17.

1.4 COMPOUND INTEREST

Suppose you were to receive a windfall of $1000. What would you do with it? While you make up your mind, let's put it in the bank for safekeeping. If you were to put the money in a time certificate of deposit for one year at 12% simple interest, then at the end of the year you would get back your original $1000 plus 12% of $1000, that is,

$$1000 + .12*1000 = 1.12*1000$$

or $1120.00. The "return" or "yield" on your investment is exactly 12%.

If, on the other hand, you were to shop around and find a bank that paid 12%, "compounded" quarterly, the result would be different. Suppose you bought a 1-year certificate of deposit under these conditions. Then after one "compounding period," in this case a quarter (of a year), you would have earned some interest—not the full 12%, to be sure, but the appropriate fraction, namely, $(1/4)*12\% = 3\%$. So, at the beginning of the second quarter, your account would already be credited with the interest earned during the first quarter, namely, 3% of $1000, or $.03*1000 = 30$. The whole point of compounding is that during the second quarter, you earn interest on the entire $1030.00.

At mid–year, the end of the second compounding period, it's time to credit your account with more interest, this time 3% of $1030. Now, $.03*1030 = 30.90$, and $30.90 added to the $1030 already in the account yields $1060.90. Moreover, the entire amount earns interest during the third quarter.

At the end of the third quarter, the interest credited to your account is $.03*1060.90 = 31.83$, rounded up. Thus, going into the fourth quarter, the amount of money you have earning interest for you is $1060.90 + $1092.73. Finally, at year's end, you stand to receive

$$1092.73 + .03*1092.73 = 1.03*1092.73$$

$$= 1125.51$$

Because of compounding, your yield is \$5.51 more than it would have been under a simple interest arrangement. Indeed, annual yield is defined in terms of the simple interest that would have returned the same amount. In this case, the annual yield on 12% compounded quarterly is 12.551%; that is, an account paying simple interest would have to offer 12.551% to produce the same total interest payment as an account paying 12% compounded quarterly. More generally, if you leave P dollars in an account for a year and get Q dollars back, the *yield* is $(Q-P)/P$, expressed as a percent.

If we can pick up an extra \$5.51 by compounding quarterly, think how much better we'll do by shopping around until we find a bank paying 12% compounded monthly! But, let's look for some short cuts before doing a lot of calculation. To simplify writing things down, let P stand for the "principal," the \$1000. Suppose r is the (yearly) rate of interest, expressed as a decimal (in our case $r = .12$), and let m denote the number of compounding periods in a year. Let's put the P dollars into the bank, sit back, and watch the action.

After the first compounding period, the amount in the account is P plus the interest on P, namely, $(r/m){*}P$. So, at the beginning of the second period, the amount of money earning interest is

$$P + (r/m){*}P = P[1+(r/m)]$$

At the end of the second period, it's time for the bank to add to the account again. During this period, $P[1+(r/m)]$ has been earning interest. Going into the third period, we have $P[1+(r/m)]$ plus the interest on $P[1+(r/m)]$, that is,

$$P[1+(r/m)] + (r/m){*}P[1+(r/m)]$$

Taking out the common factor, $P[1+(r/m)]$, we have

$$P[1+(r/m)]{*}[1+(r/m)] = P[1+(r/m)]^2$$

When interest is again posted, at the end of the third period, the account will contain a total of

$$P[1+(r/m)]^2 + (r/m){*}P[1+(r/m)]^2 = P[1+(r/m)]^2{*}[1+(r/m)]$$

$$= P[1+(r/m)]^3$$

A pattern has emerged. After x compounding periods, the total amount in the account is $P[1+(r/m)]^x$. In particular, at the end of a year, the value of the account is

$$V = P[1+(r/m)]^m \qquad (1.1)$$

Let's just confirm Equation (1.1) for the previous problem of quarterly compounding:

$$1000[1+(.12/4)]^4 = 1000[1.03]^4$$
$$= 1000((1.03) \uparrow 4)$$
$$= 1000*(1.125508\ldots)$$
$$= 1125.51$$

rounded to the nearest cent.

Now, using the computer, it's an easy matter to find the return on $1000 invested for one year at 12% compounded monthly. It's

$$1000[1+(.12/12)]^{12} = 1000[1.01]^{12}$$
$$= 1000((1.01) \uparrow 12)$$
$$= 1000*(1.126825)$$
$$= 1126.83$$

While compounding quarterly resulted in an annual yield of 12.551%, compounding monthly produces a (disappointing) yield of 12.683%, only 0.132% more. What about compounding daily?

$$1000[1+(.12/365)]^{365} = 1000[1.000328\ldots]^{365}$$
$$= 1000(1.000328\ldots) \uparrow 365$$
$$= 1000*(1.127474)$$
$$= 1127.47$$

In the course of a year, we gain an extra 64 cents by compounding daily instead of monthly. (Is it worth a drive to the next county to find a bank that compounds daily?)

In summary, the annual yield on a 12% certificate of deposit is

Number of compounding periods	Yield
1	12%
4	12.551%
12	12.683%
365	12.747%

What if you were to let your money ride, under the same interest and compounding conditions, for a total of t years? Since each of the t years has m compounding periods, the value of the account at the end of t years is

$$P[1+(r/m)]^{mt} = \text{P} * (1+(r/m)) \uparrow (m*t) \qquad (1.2)$$

To find, for example, the value of $1462.38 invested for 3 years at 9% compounded quarterly, simply ENTER

    PRINT 1462.38*(1+(.09/4))↑(4*3)

Be careful to express the interest as a decimal, and do not use the dollar sign or the comma from $1,462.38.

Exercises *(1.4)*

1. Without using the computer, determine which is bigger: $(1.1)^{10}$, $(1.01)^{100}$, $(1.001)^{1000}$, $(1.0001)^{10000}$. Justify your answer. (*Hint:* Think in terms of compound interest.)

2. Compute $(1.01)^n$ for $n = $
   a. 10   b. 100   c. 1000   d. 10000

3. Compute the final amount if $1000 is deposited for one year at
   a. 9.75% simple interest
   b. 9.5% compounded quarterly
   c. 9.25% compounded monthly
   d. 9% compounded daily

4. What is the yield for each of the deposits in Exercise 3?

5. Compute the final amount if $1000 is deposited for 5 years at
   a. 6% compounded weekly (52 times a year)
   b. 7% compounded daily
   c. 10% simple interest (i.e., compounded yearly)
   d. 8.5% compounded quarterly
   e. 17% compounded monthly

6. What is the final amount if $10,000 is deposited at
   a. 3% compounded quarterly for 30 years?
   b. 6% compounded quarterly for 15 years?
   c. 9% compounded quarterly for 10 years?
   d. 15% compounded quarterly for 6 years?

7. The "half-life" of a radioactive substance is the time it takes for half of a sample to decay. Neptunium, atomic number 93, was the first of the "man-made" elements to be discovered. The isotope Np-237 was produced in Berkeley, in 1940, by McMillan and Abelson who bombarded uranium with cyclotron produced neutrons. Another isotope, Np-241, has a half-life of 1 hour. Suppose 1 gram of Np-241 were produced in a scientific experiment. What fraction of the substance would still remain 24 hours later?

8. Suppose the population in a certain state has been growing at the constant rate of 5% every 10 years, for the last 100 years. What is the ratio of the population today to what it was a century ago?

9. If the population in a certain city is expected to grow at the rate of 1% a year for the next 20 years, compute the total growth over the next two decades. (Express the 20-year growth as a percent of the present population.)

10. Suppose inflation runs at a constant rate of 7% a year. If a "typical" item costs $100 today, what will it cost 10 years from now? (*Hint:* Think of inflation as an "interest" applied to prices. Assume the 7% is compounded yearly.)

1.5 ANNUITIES

In the last section, we discussed compound interest. If you were to deposit P dollars at a (yearly) interest rate r, compounded m times a year, the amount that you would receive at the end of t years is

$$A(t) = P(1 + r/m)^{mt} \qquad (1.3)$$

Suppose that you are going to need $1000 in 2 years for some particular purpose. How much should you deposit now so that the amount in your account 2 years from now will be $1000? What we need to do is solve Equation (1.3) for P:

$$P = A(t)/(1 + r/m)^{mt} \qquad (1.4)$$
$$= A(t)(1 + r/m)^{-mt}$$

In our case, $t = 2$ and $A(2) = 1000$. If we suppose the current interest rate on a 2-year certificate of deposit is 12%, compounded quarterly, then we can obtain P by ENTERing

    PRINT 1000*(1+.12/4)↑(−4*2)

This, of course, produces the same result as

    PRINT 1000/((1+.12/4)↑(4*2))

(namely, $789.41).

   The more usual situation is periodically to deposit smaller amounts, gradually building up to the $1000. Suppose you deposit R dollars each week for the next 104 weeks, how much should R be so that the final amount is $1000? Assume the interest rate is r and, for simplicity, that the compounding period is also weekly. Then the first R dollars will earn interest for 104 weeks and will contribute $R*(1+i)^{104}$, where $i = r/52$ is the weekly interest. The second R dollars will earn interest for 103 weeks and will contribute $R*(1+i)^{103}$ to the total, and so on. Finally, with one week to go, the last R dollars are deposited. It earns interest for only one week and contributes $R*(1+i)$ to the total. Thus, we want to solve the following equation for R:

$$1000 = R*(1+i)^{104} + R*(1+i)^{103} + \cdots + R*(1+i)$$

Factoring out the common factor, $R*(1+i)$, and writing the sum in reverse order, we obtain

$$1000 = R*(1+i)*[1 + (1+i) + (1+i)^2 + \cdots + (1+i)^{103}] \quad (1.5)$$

   As we're about to do some algebraic manipulations, it will simplify things to temporarily use x to denote $(1+i)$. The difficult part of Equation (1.5) involves the evaluation of a sum like

$$S = 1 + x + x^2 + \cdots + x^{n-1} \quad\quad\quad (1.6)$$

where, in our case, $n = 104$. There is an algebraic trick that allows us to obtain S without doing lots of adding. Just notice

$$xS = x + x^2 + x^3 + \cdots + x^n$$

$$= S - 1 + x^n$$

so that

$$(x - 1)S = (-1 + x^n)$$

and

$$S = (x^n - 1)/(x - 1) \quad\quad\quad (1.7)$$

Ours is the situation in which $1000 = RxS$. So, $R = 1000/xS$. Using Equation (1.7),

$$R = 1000(x - 1)/[x(x^n - 1)]$$

It only remains to remember that $x = 1 + i$, so that

$$R = 1000i/[(1 + i)((1 + i)^n - 1)]$$

Suppose the current passbook savings account pays 6%, compounded weekly. Then to save the amount $A = \$1000$ by depositing R dollars a week, R can be obtained by ENTERing this sequence of instructions:

```
I  =  . 0 6 / 5 2
X  =  1 + I
R  =  1 0 0 0 * I / ( X * ( X ↑ 1 0 4 − 1 ) )
PRINT  R
```

The result is $R = $ just under $9.05. It is worth observing that this sequence of instructions is very close to a computer program.

Before going on, let's write down the general formula. Suppose the interest rate you are able to obtain is r, compounded m times a year. Assume that you deposit R dollars at the beginning of each compounding period with the goal of reaching A dollars at the end of t years. Then

$$R = Ai/[(1+i)((1+i)^{mt} - 1)], \qquad (1.8)$$

where $i = r/m$ is the interest per compounding period (and t can be any fraction of a year that is a whole number of compounding periods).

Let's do another example. Suppose that, 6 months after your Aunt Mary's 50th birthday, she starts to prepare seriously for her retirement. She wants to begin making monthly deposits so that when she retires at age 65, she will have $50,000 in the bank to augment her retirement income. What should her monthly deposit be if her credit union pays 8.5%, compounded monthly? To find out, ENTER this sequence of instructions based on Equation (1.8):

```
A = 5 0 0 0 0
T = 1 4 . 5
M = 1 2
I = . 0 8 5 / M
X = 1 + I
R = A * I / ( X * ( X ↑ ( M * T )  −  1 ) )
PRINT  R
```

Up to now, the examples all concern people with the foresight and discipline to save toward a goal. On the other hand, the unexpected can happen to any of us. Suppose that while Aunt Mary is working out the above calculation in her head, she neglects to watch where she is going and drives her car into a tree. With the insurance settlement in her pocket ($4500), she goes shopping for a new car. Finding one she likes for $7500, Aunt Mary uses the settlement money for a down payment and decides to finance the remaining $3000 over the next 3 years. What will her monthly payment be?

Let's work the problem out in general and then make substitutions for Aunt Mary's particular situation. Suppose you borrow a sum of money, M. Let's assume the (yearly) interest rate is r, compounded m times a year, coinciding with your payment. What should the payment be in order to retire the debt in t years? At the end of the first period after the money is borrowed, the bank adds interest on the total amount M, and you pay P. Thus, at the beginning of the second compounding period, you owe $M(1+i) - P$, where $i = r/m$. At the end of the second period, the bank assesses interest on this amount, and you pay another P dollars. Thus, at the beginning of third period, you owe the bank

$$[M(1+i) - P](1+i) - P$$

At the end of the third period, the amount still owed (on which interest will be computed at the end of the fourth period) is

$$([M(1+i) - P](1+i) - P)(1+i) - P$$

If we continue in this way until the end of the nth period, at which time we want the loan to be paid off, we find (after some rearrangement of the terms) that

$$0 = M(1+i)^n - P[(1+i)^{n-1} + (1+i)^{n-2} + \cdots + 1]$$

But, this just brings us back to Equations (1.6) and (1.7) again. In particular, we can conclude that

$$P[(1+i)^n - 1]/i = M(1+i)^n$$

so that

$$P = Mi(1+i)^n/[(1+i)^n - 1] \qquad (1.9)$$
$$= Mi/[1 - 1/(1+i)^n]$$

In Aunt Mary's case, $M = 3000$. Since $t = 3$ years, and we're making monthly payments, $n = 3*12 = 36$. If the interest rate is 16%, compounded monthly, then $i = .16/12$. The monthly payment P can be obtained from Equation (1.9) by ENTERing the sequence

```
M=3000
N=36
I=.16/12
P=M*I/(1-1/(1+I)↑N)
PRINT P
```

Exercises (1.5)

1. Suppose you are able to earn 8% interest, compounded monthly. How much should you deposit now in order to have $1000 in your account n years from now if n is
   a. 1   b. 2   c. 3   d. 4   e. 5

2. Suppose you are able to earn an interest rate r, compounded monthly. How much should you deposit now in order to have $1000 in your account 2 years from now if r is
   a. 7%   b. 8%   c. 9%   d. 10%   e. 11%   f. 12%

3. Show that
   a. $1 + 2 + 4 + 8 + \cdots + 2^{19} = 2^{20} - 1$
   b. $1 + 1/2 + 1/4 + \cdots + (1/2)^{19} = 2 - (1/2)^{19}$

4. How much does Aunt Mary have to deposit each month to reach her goal of $50,000 at age 65?

5. Suppose it is your goal to save $5000 over the next 4 years by depositing R dollars each month in a savings account which pays an interest rate r, compounded monthly. What does R have to be if r is
   a. 8%   b. 9%   c. 10%   d. 11%   e. 12%

6. What will Aunt Mary's (monthly) car payment be?

7. What would Aunt Mary's (monthly) car payment have been had she shopped around for a loan at 14.5%? How much money would this have saved her over the 3-year life of the loan?

8. Suppose you went to a relative for help in financing the purchase of a used car priced at $5000. Assume the going rate for used car loans is 14.5%, compounded monthly. Suppose your relative offers you a choice between
   a. an outright gift of $500 to be used as a down payment, or

    b. a personal loan of the full $5000 at 8%, compounded monthly.

    Assuming, in either case, that you plan to pay off the loan in three years, which option costs you less over that period?

9. Suppose you decide to buy a house. What will your monthly payments be if you take out a fixed-rate, 30-year, $90,000 mortgage at the following rates, compounded monthly:

    a. 9%   b. 10%   c. 11%   d. 12%   e. 13%   f. 14%

10. What is the total amount paid to the bank over the 30-year life of the mortgage in (each part of) Exercise 9.

Chapter 2
COMPUTER PROGRAMMING

2.1 LINE NUMBERS

It's time to take the plunge into programming. Up to now, we have been communicating directly with the computer in what is sometimes called the "interactive mode." In this mode, the computer responds to single instructions.

A program is a sequence of instructions that the computer executes in a prescribed order. In the BASIC[†] language we prescribe the order by means of line numbers. Each instruction is given a number. Unless directed otherwise, the computer executes its instructions in numerical order. ENTER the following line exactly as it appears here.

```
10 PRINT 3*5
```

The computer did not respond because the line number "10" indicated that this is a line in a program. In this case, the ENTER key instructed the computer to remember this line. You can test its memory by ENTERing the BASIC instruction

```
LIST
```

(That is, type the word "LIST" and press the ENTER key.) To make the computer execute this program, ENTER the command

```
RUN
```

[†]Beginners All-purpose Symbolic Instruction Code.

Once executed, the program is retained in the computer's memory. Confirm this by ENTERing the LIST command. Indeed, the computer will RUN the program again, should you so desire.

Let's try another program. ENTER these two lines:

```
10  A=3*5
20  PRINT  A
```

The first of these instructions does two things. Since we have reused line number 10, the computer has replaced the old line 10 in its memory with the new line 10. Confirm this by ENTERing the LIST instruction. Then try this alternative

```
LIST  10
```

When you RUN the program, the response will be exactly the same as before. But, internally, there is a difference. The computer first executes the instruction in line 10, that is, it computes 3*5 and remembers the answer by putting "15" in a memory location which it then labels "A." Having executed the lowest numbered line in the program, the computer looks for the next higher number. In this case, it finds line 20 and executes that line by PRINTing the contents of memory location A (but forgetting that the contents of that location represents 3*5). Finding no line number higher than 20, it awaits further instructions.†

The RUN command has another feature that is sometimes good and sometimes bad. When RUN is ENTERed, memory locations like the one labeled "A" are cleared (i.e., emptied or "forgotton"). Of course, if you RUN this same program again, "15" will immediately be put in a (generally different) memory location labeled "A." ENTER this (directly, without a line number)

```
PRINT  A
```

Now try this instruction (again, in the interactive mode, without a line number)

```
A  =  A+3
```

---

†In some older versions of BASIC, it is necessary to indicate explicitly to the computer that the program has ENDed. This is done by adding another line, e.g., 30 END. While this instruction is no longer needed, it still represents good programming technique.

This may seem like a strange thing to type, until we remember that the computer does not take it as an algebraic statement. (In particular, the machine will not break itself trying to make 15 equal to 18.) To the computer, the phrase "A = A + 3" is taken as an instruction to change the contents of memory location A. Specifically, the computer replaces the "15" currently residing in location A with the new value 15 + 3 (= 18). Confirm this by ENTERing

    PRINT A

Before we can experiment with other programs, we need to get rid of the program currently being held in memory. One way to do that is to re-use line numbers 10 and 20 (and, perhaps 30). Alternatively, by typing a line number, say 20, followed by ENTER, we can erase (or delete) line 20. Try this and confirm the result by ENTERing LIST.

Of course, it would be nice to have a simple way to instruct the computer to dump (i.e., erase or forget) the old program. This is done by ENTERing the BASIC instruction

    NEW

Let's finish by writing a program to add any two numbers, without knowing in advance what they might be. This involves one more word in our growing BASIC vocabulary! The INPUT instruction. When the computer encounters the INPUT command, it pauses to receive "input." ENTER these lines

    20 INPUT A
    30 INPUT B
    50 PRINT A+B
    60 END

When you ENTER the RUN instruction, the computer will come to line 20 and wait for a number. It will indicate that it expects something from you by PRINTing a question mark. Respond to this "prompt" by ENTERing a number, say 3.14. (That is, type 3.14 and press the EN-TER key.) Having received your "input" and put it in a memory location that it calls "A," the computer proceeds to the next highest numbered line, in this case, 30. Since this is also an INPUT instruction, expect to see another prompt, i.e., another "?". Respond by ENTERing another number, say 1.23. The computer will accept this input and put it in a memory location labeled B. Finally, in response to line 50, the computer

PRINTs 4.37, the sum of the numbers it "recalls" from memory locations A and B. Run this program a few times with numbers of your own choosing. (At least once, choose a negative number for B.)

Let's put a little "class" into the program. (Since we'll only be adding to the program, not starting over, do not ENTER the NEW instruction.) Type and ENTER these lines

```
10  PRINT  "ENTER  TWO  NUMBERS"
40  PRINT  A ; "+" ; B ; "=" ;
```

Now, LIST the program again. Notice that the computer automatically puts the lines in numerical order, regardless of the order in which they were originally ENTERed.

See if you can anticipate the output before ENTERing RUN. (The first three semicolons in line 40 may not be needed on your machine. The last one will be. It is a way of telling the computer not to begin a new line when it reaches the next PRINT statement, but rather to continue from where it left off with this PRINT statement. Experiment by deleting some or all of the semicolons from line 40.†

Exercises   (2.1)

In all the exercises that call for you to write a program, write one with some "style," that is, use one or more PRINT commands similar to the one in line 10 of the last program above.

1. Write a program to INPUT two numbers and PRINT their product.

2. Write a program to INPUT three numbers and PRINT their sum.

3. Write a program to INPUT two numbers and PRINT their quotient. RUN your program and see what happens when you try to divide by zero.

4. Write a program to INPUT a number N and PRINT 2 ↑ N.

5. Try out the program you wrote in Exercise 4. Use it to check some of the numbers you tabulated in Exercise 10, Section 1.3.

---

†To do this, you will need to reENTER (a new) line 40. In a five-line program, it is rarely good technique to use line numbers 1–5. It is a rare programmer who won't want to make changes as he/she goes along. Generally, it is a good idea to leave gaps of at least 10 between successive line numbers.

6. Write a program to INPUT two numbers, say M and N. Have the computer output 2^M, 2^N, $(2^M)*(2^N)$, and 2^{M+N}.

7. RUN the program you wrote in Exercise 6 for the following pairs M, N and record the responses.
   a. 5,6   b. 7,9   c. 12,6   d. 0,0   e. 4,7   f. 19,9

8. What is 2^0? (*Hint:* See Exercise 7d.)

9. Explain why $(2^M)*(2^N) = 2^{M+N}$. (*Hint:* This is a question in mathematics, not computing. You may find your tabulation from Exercise 10, Section 1.3 to be helpful. Another helpful idea would be to work out the solutions to Exercise 7 by hand. You concentrate on the process and use the computer's responses to check your arithmetic.)

10. RUN the program you wrote in Exercise 6 for the following pairs M, N and record the responses.
    a. $5, -3$
    b. $9, -5$
    c. $6, -7$
    d. $-3, 5$
    e. $-1, -1$
    f. $0, -1$

11. Explain (mathematically) the responses given by the computer in Exercise 10.

12. Tabulate $2^{-1}, 2^{-2}, 2^{-3}, \ldots, 2^{-20}$. (*Hint:* $2^{-1} = 1/2 = 0.5, 2^{-2} = 1/2^2 = 0.25, 2^{-3} = 1/2^3 = 1/8 = 0.125, \ldots$ Of course, if you use the computer for these computations, it will PRINT, e.g., .25 and not 0.25. In this text, the superfluous "0" is used to emphasize the position of the decimal point.)

13. Repeat Exercises 6–11 with "2" replaced by "3."

14. Repeat Exercise 6 with "2" replaced by a number A which is to be INPUTted.

15. Write a program which INPUTs the temperature in degrees Fahrenheit and PRINTs the temperature in degrees centigrade. (*Hint:* See Exercise 7, Section 1.2.)

16. Write a program which INPUTs the temperature in degrees centigrade and PRINTs the temperature in degrees Fahrenheit. (*Hint:* See Exercises 9–10, Section 1.2.)

17. Write a program to INPUT a decimal number N and PRINT N as a percent.

2.2 LOOPS

The BASIC language has been around long enough to have evolved some abbreviations, colloquialisms, and even some different dialects.[†] In most of the dialects, "?" is an abbreviation for "PRINT". To see if this abbreviation works on your machine, try ENTERing

```
? "IT WORKS!"
```

If your computer responded with the SYNTAX ERROR message, then it doesn't work. Another short-cut is to INPUT more than one variable at a time. Test this variation with the following simple program.

```
10  INPUT X,Y
20  PRINT "NO PROBLEM."
30  PRINT "X =" X
40  PRINT "Y =" Y
50  END
```

When you RUN the program, you will get a single prompt. Respond to it by typing in two numbers separated by a comma, e.g., 8,34. If you get an error message, LIST the program and check it carefully for typographical errors. If you get an EXTRA IGNORED message, it may mean that your version of BASIC does not "support" double INPUT-ting.

Another variation of the INPUT command involves a sort of combination with the PRINT instruction. ENTER the following line

```
10  INPUT "ENTER TWO NUMBERS";X,Y
```

LIST the program to be sure that the new line 10 has been substituted for the old line 10 and that the rest of the program is intact. Then RUN the revised program.

---

† The BASIC used in this book is valid in all the common dialects. Whenever you have problems, look for (1) typing mistakes, (2) errors of punctuation, (3) logical errors in your program, and then (4) variations in the BASIC dialect.

Even if the BASIC dialect "spoken" by your computer supports this modification of the INPUT statement, it may PRINT a grammatically incorrect question mark. Keep that in mind when you use this variation so you can phrase your "prompts" appropriately. (If everything works as expected, try the modified INPUT statement without the semicolon. Does it work?)

As we have seen, a computer executes the sequence of instructions constituting a program in line-number order, regardless of the order in which the lines were originally typed in. There is a way to direct the computer to a line other than the one with the next higher number. This involves the GO TO command. Consider (just consider, don't type it in or RUN it yet) the following program:

```
10 PRINT "[YOUR NAME]"
20 GO TO 10
30 END
```

A computer RUNning this program will first encounter line 10. It will print your name and look for the next higher numbered line, which is 20. At line 20, it will follow the instruction to GO TO 10, and just repeat the same process, over and over again. This kind of construction is called a "loop," in this case, an infinite loop! The computer will never get to line 30 and stop. This suggests the following question: "Is there any way to interrupt a computer while its attention is engaged by the execution of a program?" The answer is "Yes," but the exact mechanism by which this is done varies from one machine to the next. On some computers, e.g., the Commodore 64, there is a "RUN/STOP" key. On others, the interrupt function is accomplished by holding down the CTRL key and simultaneously pressing another key. (On the Apple II, the second key is "C," on the IBM PC, it is the "scroll lock.")

After you have determined the interrupt procedure for your machine, test it by RUNning the program above. (Don't forget to first ENTER the NEW command.)

One difference between BASIC dialects is the attention given to spaces between code words. For example, in some versions of BASIC, the GO TO command may be written GOTO, without a space. Other versions will respond to GOTO with the SYNTAX ERROR message. Try replacing line 20 with

```
20 GOTO 10
```

and see which variation your computer uses.

Another way to deviate from the usual order of program execution is by means of the IF ... THEN instruction. This is a powerful and versatile command. Consider the following program.

```
20  N=0
30  PRINT  "[YOUR  NAME]"
40  N=N+1
50  IF  N<10  THEN  30
60  END
```

Let's analyze the program line by line. In line 20, we instruct the computer to store the number "0" in a memory location labeled N. (This must be done within the program if we want it done at all. If we ENTER N = 0 in the direct mode, and then ENTER the RUN command, the computer will delete the "0" from location N and will eliminate the label N itself.) In line 30, we instruct the computer to PRINT your name. Line 40 changes the number in memory location N. (The new content of that location being one more than the old content.) Line 50 constitutes a fork in the road. If the current content of memory location N is less than 10, the computer is directed back to line 30. Otherwise (if the content of N is not less than 10), the computer goes on to line 60.

Logically, N serves as a counter. The number in memory location N is the number of times your name has been printed. In line 50, we effectively ask the computer if your name has yet been printed 10 times. If it has, then END the program. If it has not, then go through the loop one more time. ENTER the (NEW) program and RUN it.

Maybe you would like your name PRINTed 12 times, or maybe just 5 times. Let's modify the program to accommodate you. ENTER these lines

```
10  INPUT  "HOW  MANY  REPETITIONS";R
50  IF  N<R  THEN  30
```

When you RUN the program, does a question mark appear in the opening prompt? (Make a mental note so you will know whether or not to supply one the next time a similar situation arises. If your BASIC dialect does not support the modified INPUT statement, make the question a PRINT instruction in line 9, and replace line 10 with INPUT R.)

Exercises   (2.2)

1. Describe the "interrupt procedure" for breaking in while the computer is RUNning a program.

2. RUN this program (with the semicolon in line 10) and describe how the response differs from the corresponding program in the text.

```
10 PRINT "[YOUR NAME]";
20 GO TO 10
```

(Did you remember that you can abbreviate PRINT?)

3. RUN this program (with the comma) and describe how the response differs from what you observed in Exercise 2.

```
10 PRINT "[YOUR NAME]",
20 GO TO 10
```

(Did you think of ENTERing NEW?)

4. RUN the program you wrote for Exercise 3 and interrupt it. Then ENTER the following instruction

```
CONT
```

Describe what happened.

5. John Wallis (1616–1703) was the leading English mathematician before Isaac Newton. In his treatise *Arithmetica infinitorum* (published in 1655), Wallis presented his famous "infinite product" formula

$$\frac{\pi}{2} = \frac{2*2*4*4*6*6*8*\cdots}{1*3*3*5*5*7*7*\cdots} = \frac{2*2}{1*3} * \frac{4*4}{3*5} * \cdots$$

The dots mean that the indicated pattern continues on forever. Since even the computer cannot be expected to "continue forever," what good is such an expression? The answer is that by terminating both the top and bottom at the same place, we can obtain an approximation for $\pi/2$, and, hence, for π itself. See what kind of an approximation is given if the expression is terminated just before the dots, that is, compute

$$2 * \frac{2*2*4*4*6*6*8}{1*3*3*5*5*7*7}$$

and see how close it is to π.

6. As you saw in Exercise 5, the approximation to π produced by using seven terms of Wallis's infinite product is off by more than 0.2. The following program produces the approximation to π using the first $2N$ terms of Wallis's expression. Write out a line by line explanation of the program.

```
10  PRINT "INPUT A POSITIVE INTEGER."
20  INPUT N
30  P=2
40  I=0
50  I=I+1
60  P=P*(2*I)↑2/((2*I—1)*(2*I+1))
70  IF I<N THEN 50
80  PRINT "THE" 2*N "-TH ESTIMATE IS" P
90  END
```

7. Type in the program given in Exercise 6. RUN it and record the response for N = 50, 100, 200, 300, and 1000. Determine as accurately as you can how long the computer takes for each of these RUNs.

8. Write a program to compute the sum, $1 + 2 + 3 + \cdots + 99 + 100$, of the first 100 positive integers.

9. Write a program to INPUT a positive integer, N, and then to compute the sum, $1 + 3 + 5 + \cdots + (2N - 1)$, of the first N odd positive integers. (*Hint:* The Nth odd positive integer is $2N - 1$.)

10. Explain the relationship between Exercise 9 above, and Exercises 2 and 5 of Section 1.3.

11. There is a story told that the "Commander of the Faithful" was so pleased with the new game of Chess that he invited its inventor to name his own reward. As the chessboard was on the table before the monarch, the inventor asked for 1 grain of wheat to be placed on the first square of the board, 2 grains on the second square, 4 on the third, and so on, each succeeding square getting twice the number of grains as the previous one. The reward claimed by the inventor was the total number of grains on the board after the

64 th square had received its allotment. Pleased by what he thought was a modest request, the potentate sent for a bag of grain while he contemplated what further reward he might bestow on this man who had pleased him twice. (What further reward do you think was eventually bestowed on the inventor?) Write a computer program to determine how many grains of wheat were demanded by the inventor.

12. RUN your program from Exercise 11, and compare the answer with 2^{64}, the number of grains which would have been placed on the 65th square, had the pattern continued beyond the 64 squares of the chessboard.

13. Without ENTERing this program, explain what it would do were it to be RUN.

```
 10 GO TO 90
 20 PRINT "ELLO";
 30 PRINT " ";
 40 GO TO 70
 50 PRINT "MAN"
 60 GO TO 110
 70 PRINT "HU";
 80 GO TO 50
 90 PRINT "H";
100 GO TO 20
110 END
```

2.3 NUMBER SEQUENCES

Some standardized I.Q. tests involve the ability to recognize patterns such as those in sequences of numbers. Can you tell what the next number is in each of the following sequences?

(1)   3, 7, 11, 15, ___
(2)   4, 7, 10, 13, ___
(3)   3, 6, 12, 24, ___
(4)   2, 3, 5, 7, ___
(5)   1, 1, 2, 3, ___
(6)   12, 24, 36, 48, ___
(7)   2, 4, 8, 16, ___

One problem with tests of this type is that there may be more than one "right" answer. Take the 6th sequence for example. Is the next number 60? Are you sure? Would you stake your life on it? Suppose I were just recording the numbers that appear on my digital stopwatch at 12-second intervals? Who is right?

In (4), I had in mind the sequence of prime numbers, so that the next number is 11. But, couldn't one present an argument for the next number being 9, say? Suppose a few more numbers are listed for sequence (5):

$$(5') \quad 1, \quad 1, \quad 2, \quad 3, \quad 5, \quad 8, \quad 13, \quad 21, \quad -$$

(I have 34 in mind for the next number.)

Consider sequence (1). Let's agree, not only that the next number is 19, but that each successive number after that is obtained by adding 4 to the previous number. Then we have agreed to an *algorithm* for producing each succeeding number (and eliminated all doubt about what it is). This particular algorithm (or recipe) generates a so-called arithmetic sequence. An arithmetic sequence is one that begins with an arbitrary number, call it N, and involves a second number, say A. (It may happen that A and N are the same.) Each succeeding number in the sequence is then obtained by adding A to the previous number. In (1), $N = 3$ and $A = 4$. In (2), $N = 4$ and $A = 3$.

Returning to (1), suppose we're asked, not for the 5th, but for the 100th number in the sequence? One way to approach this problem is to convert the algorithm into a computer program and RUN the program. Let's do it. Here is one possibility:

```
10  T = 3
20  I = 1
30  I = I + 1
40  T = T + 4
50  IF  I < 100  THEN  30
60  PRINT  "THE  100 - TH  NUMBER  IS" ; T
70  END
```

This is another example of a program with a loop. In fact, this kind of looping occurs so frequently that there is a specific BASIC simplification for it, namely, the FOR . . . NEXT construction. We can illustrate it by replacing lines 20–30 in our program with the single line

```
20  FOR  I = 2  TO  100
```

and line 50 with

```
50 NEXT I
```

The new lines 20 and 50 set up a loop. The logic is as follows: The first time the computer encounters (the new) line 20, it sets I = 2. It then proceeds through the program until it encounters the NEXT statement. It now asks itself whether I = 100. The answer being no, control is returned to the FOR statement (in our case this occurs at line 20) where I is "incremented," that is, 1 is added to the current value of I. Using the value I = 3, the computer proceeds to the next line of the program (namely, line 40—line 30 having been eliminated), and so on, executing its instructions in line number order, until it again reaches the NEXT statement. Once again, the machine asks itself whether I = 100. The answer again being no, control is cycled back to the FOR statement in line 20, where I becomes 4. This process continues, over and over, with the value of I increasing by 1 each time. Ultimately, the machine arrives at line 50 with the value of I equal to 100. Then, and only then, the execution moves on to line 60. Try RUNning both versions of the program and confirm that they give the same answer. (Don't forget to eliminate line 30 before you RUN the second program.)

Let's experiment with another program using FOR . . . NEXT. EN-TER this one, but try to figure out how the computer will respond before you RUN it.

```
100 FOR I=0 TO 9
200 PRINT I;
300 NEXT I
400 PRINT "END"
500 PRINT I
600 END
```

After you have RUN the program, try it with the following replacement:

```
100 FOR I=−5 TO 4
```

Suppose we just want to PRINT even numbers. One way to do it would be to replace line 200 with PRINT 2*I. Another way involves a variation of FOR . . . NEXT. Replace line 100 with

```
100 FOR I=0 TO 8 STEP 2
```

The "STEP 2" informs the computer that we want it to count by two's.

RUN this variation, then try substituting

```
100 FOR I=8 TO 0 STEP -2
```

It may be that we don't want to increment by whole numbers at all, but by fractions. Try this

```
100 FOR I=7.5 TO 8.3 STEP .1
```

(In the absence of "STEP", the computer assumes the "default" value of "STEP 1".)

One interesting thing about all this is that we don't need a computer to determine the 100th term in sequence (1). Let's examine our algorithm, but without doing any arithmetic. (Sometimes, doing the arithmetic obscures what is really going on.) The first number, let's call it $T(1)$, is 3. The second, say $T(2)$, is $3 + 4$. A few more are

$$T(3) = 3 + 4 + 4,$$

$$T(4) = 3 + 4 + 4 + 4,$$

$$T(5) = 3 + 4 + 4 + 4 + 4$$

and so on. What we are trying to do is discover a pattern. And, one is certainly emerging. Each term consists of 3 plus a sum of several 4's. How many 4's? Notice that $T(I)$ involves a total of I terms, one 3 and $(I-1)$ 4's. Thus,

$$T(I) = 3 + (I-1)*4$$

In particular, the 100th term in the sequence is

$$T(100) = 3 + (99)*4 = 399$$

It seems a little foolish, in hindsight, to have gone to all the trouble of writing these programs when the answer is so easy to obtain directly! There is a moral in all this. We should always keep in mind that a computer is no substitute for thinking.

The notation $T(I)$ was used in the preceding discussion in place of the more common subscripted expression T_I. The reason for this is that the BASIC way to write a subscripted variable is $T(I)$. (There is no more direct way to ENTER a subscript into the computer.)

When the computer encounters an expression of the form $T(I)$, it automatically reserves (exactly) 11 memory locations for subscripted "T" variables, namely, $T(0)$, $T(1)$, $T(2)$, ..., $T(10)$. A difficulty arises when we want to discuss $T(11)$, $T(12)$, etc. What we have to do in this situation is explictly tell the computer to set aside more room for T variables. This is done by means of a DIMension statement. As an example, let's have the computer actually PRINT out the first 100 numbers in sequence (1). ENTER and RUN this program:

```
10  DIM  T(100)
20  T(1)=3
30  PRINT  T(1);
40  FOR  I=2  TO  100
50  T(I)=T(I-1)+4
60  PRINT  T(I);
70  NEXT  I
80  END
```

Line 10 shows the format for the DIMension statement. The computer is being instructed to reserve space for 101 entries labeled T(0) up to T(100). (In this particular case, we aren't going to use the memory location labeled T(0). It will simply hold the "default value," 0. In general, we are not obliged to use all the reserved memory locations, but we can't use more locations than have been reserved.† Lines 20 and 30 start the ball rolling. The semicolon in line 30 tells the computer to PRINT the succeeding number on the same (screen) line. (Program) line 40 begins the loop and indicates its length. Line 50 computes the next number in the sequence and line 60 PRINTs it. Finally, line 70 is the bottom of the loop. Try RUNning the program.

Notice that the computer doesn't have sense enough to move down to the next line when there isn't enough room to PRINT an entire number at the end of the line. In fact, that's our fault. The computer is doing exactly what the semicolons instructed it to do. Try replacing the semicolons in lines 30 and 60 with commas, and RUN the program again.

There is an annoying technical difficulty involved with the use of DIMension statements. A REDIM'D ARRAY ERROR may occur if the computer encounters a second DIM statement for the same variable. (This happens, for instance, if the DIM statement occurs within a loop.) Try RUNning this program:

---

† We could always "play it safe" and reserve (lots) more memory locations than we expect to use. This is fine as long as it is understood that any computer's memory is finite. A 64K RAM microcomputer, for example, has 64,000 memory locations, not all of which are "user accessible."

```
10  DIM  T(100)
20  DIM  T(150)
30  END
```

Notice, on the other hand, that there is no difficulty with this one:

```
10  DIM  T(100)
20  DIM  U(150)
30  END
```

It is good programming practice to put all the DIMension statements near the beginning of the program. In any case, any DIM statement should appear before the first use of the corresponding variable.

Exercises   (2.3)

1. Rewrite the program in Exercise 6, Section 2.2, using the FOR ... NEXT construction to generate the loop.

2. Rewrite your solution to Exercise 8, Section 2.2, using FOR ... NEXT.

3. Find the 100th term in the arithmetic sequence 1, 6, ... without using the computer.

4. Find the 673-rd term in the arithmetic sequence 19, 21, ... without using the computer.

5. Sequence (3) in the text is a so-called *geometric* sequence. A geometric sequence begins with a number, call it N, and involves a second number, say A, such that each successive number is the product of A and the previous number. In sequence (3), $N = 3$ and $A = 2$. In (the familiar) sequence (7), both N and A are 2. Consider the following sequences. Determine which are arithmetic and which are geometric. For the arithmetic/geometric sequences, determine both N and A. If a sequence is neither arithmetic nor geometric, try to give an algorithm for producing successive terms.
   a. 1, 4, 9, 16, 25, ...
   b. 1, 2, 6, 24, 120, ...
   c. 1, 2, 3, 4, 5, ...
   d. 1, 0, 0, 0, 0, 0, ...
   e. 7, 11, 15, 19, 23, ...
   f. 7, 49, 343, 2301, 16807, ...

g. 2, 6, 18, 54, 162, ...

h. 1, 2, 5, 10, 20, ...

i. 3, 11, 19, 27, 35, ...

j. 1, 1, 2, 3, 5, 8 ...

6. Given a geometric sequence (see Exercise 5) with first number N and multiplier A, find a way to produce the k-th term in the sequence without having to determine the preceding $k-1$ terms. That is, find an explicit formula, in terms of k, for the k-th term in the sequence.

7. The numbers that occur in sequence (5)–(5′) and in Exercise 5j are called "Fibonacci" numbers. The following is a program to produce the k-th Fibonacci number. Do a line-by-line analysis of the program describing the purpose of each line. (*Warning:* Some versions of BASIC do not "support" the statement in line 50. If yours is one of these, you will have to "play it safe," and replace line 50 with DIM F(1000).)

```
 10  PRINT "TO PRODUCE THE K-TH FIBONACCI"
 20  PRINT "NUMBER, ENTER K (<1000)."
 30  INPUT K
 40  IF K>999 THEN 140
 50  DIM F(K)
 60  F(1)=1
 70  F(2)=1
 80  IF K<3 THEN 120
 90  FOR I=3 TO K
100  F(I)=F(I-1)+F(I-2)
110  NEXT I
120  PRINT "THE REQUIRED NUMBER IS:"F(K)
130  GO TO 150
140  PRINT "YOU'RE IMPOSSIBLE!"
150  END
```

8. Write a program to compute and PRINT the first 20 Fibonacci numbers. (See Exercise 7.)

9. Write a program to PRINT the sum of the first 10 Fibonacci numbers. (See Exercise 7.) The sum is an exact multiple of which Fibonacci number?

10. Describe an algorithm for producing successive terms of the sequence 1, 3, 6, 10, 15, 21, 28, ...

11. Using the algorithm you devised in Exercise 10, write a program to compute the 100th term in the sequence. Then compare the program with your solution to Exercise 2.

12. Find an explicit formula for the k-th term in the sequence of Exercise 10; that is, express the k-th term as a function only of k (and not of the preceding $k - 1$ terms).

13. The k-th term in the sequence 1, 2, 6, 24, 120, . . . (that is, the product of the first k numbers) has a special notation. It is "$k!$". This expression is read "k-factorial." The exclamation mark is used only because it is a convenient symbol on a standard typewriter. In particular, $k!$ is not "k" said with special emphasis. Rather,

$$k! = k*(k-1)*(k-2)* \cdots *3*2*1.$$

Write a program to compute and PRINT $k!$ for $k = 1, 2, \ldots, 15$.

14. The Lucas sequence[†] begins 1, 3, ... The algorithm for producing more terms is the same as that for the Fibonacci numbers, that is, each successive term is the sum of the previous two. Write a program to compute and PRINT the first 50 Lucas numbers.

2.4 FLOWCHARTS

Before beginning a major project of any kind, it is frequently useful to do some preliminary thinking—to draw up some plan of attack. Computer programming falls into this category. Plans for computer programs are called "flowcharts." These constitute schematic pictures of the algorithms that we intend to implement in the program. Flowcharts

---

[†] Named after Edouard Lucas (1842–1891). We will have more to say about Lucas in Section 6.6.

help break a big job down into many small jobs. We'll use the first program of Section 2.3 (reproduced below) as an illustration.

```
10  T=3
20  I=1
30  I=I+1
40  T=T+4
50  IF  I<100  THEN  30
60  PRINT  "THE  100-TH  NUMBER  IS";T
70  END
```

Like a mall directory, a flowchart begins with a "you are here" or "start" notation, and then leads us through the logical process step by step. The first thing we want to do is "initialize" our variables, namely, tell the computer to begin with the first $(I=1)$ term which is $T=3$. In the program, this is done in lines 10 and 20. In the flowchart below, this step is roughly indicated by the "initialize" box. Having agreed with the computer on the first term in the sequence, we want it to compute what numerical place in the sequence the next term will occupy. Since the variable I is keeping track of where we are in the sequence, the next term will be the $(I+1)$st. In the program, this is handled in line 30. In the flowchart, this step is indicated by the "increment" box. Having decided the numerical position of the next term in the sequence, we instruct the computer how to determine what it is. This is done in line 40 and the "calculate next term" box. We now come to the interesting part of the program—the decision-making part. Is the term that has just been computed the 100th? If it is, we PRINT it out and END the program. If not, we cycle back to the "increment" stage and determine the numerical place in the sequence occupied by the next term to be computed.

We will consistently use ovals to start and END flowcharts. Rectangles will denote internal computer processing, diamonds will be reserved for decisions, and parallelograms (which are neither diamonds nor rectangles) will indicate input/output. Each line should be labeled with an arrow head to indicate the direction of flow.

The obvious question is, "Is it worth all the trouble?" The answer is: "Yes, sometimes." In the case of this simple program, probably not. In the case of more sophisticated programs, it may well be worth the trouble. Let's try another example, but this time we'll write the flow-chart first as an aid to writing the program (which, after all, is the whole point)!

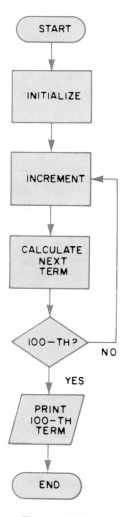

Figure   2.1

One of the most "artistic" numbers in mathematics is the so-called "golden ratio," ϕ.[†] From (at least) the time of the ancient Greeks, it has generally been agreed that there is a single most visually pleasing shape for a rectangle to have. The square is, somehow, unimaginative—even dull. (You wouldn't want to be known as a "square" would

---

[†] ϕ is the Greek letter *phi*. Like pi, the number was thought important enough by the ancient Greeks to have a letter of the alphabet reserved for its (not quite exclusive) use.

you?) The most pleasing rectangle should be longer than it is wide, but not too long. In short (so to speak), the most pleasing rectangle has ratio of length to width equal to ϕ. An ordinary 3-by-5 card is close to the "golden proportion," as are many playing cards, cereal boxes, etc. The golden ratio is approximately 1.6. It turns out that you can achieve better and better approximations to ϕ by taking ratios of successive Fibonacci numbers! That is, the sequence

$$1/1, \ 2/1, \ 3/2, \ 5/3, \ 8/5, \ 13/8, \ \ldots$$

gets closer and closer to ϕ, the more terms we compute. Let's write a program to INPUT K and compute $F(K)/F(K-1)$. But first, we'll outline the major steps by means of a flowchart.

The first couple of steps are almost automatic. (See Figure 2.2.) The main part of the program is diagrammed in Figure 2.3. It remains only to assemble the pieces (add some class), and write the program itself:

```
 10  PRINT "I WILL ESTIMATE PHI BY"
 20  PRINT "THE K-TH FIBONACCI RATIO."
 30  INPUT "WHAT IS K";K
 50  DIM F(K)
 60  F(1)=1
 70  F(2)=1
 80  FOR I=3 TO K
 90  F(I)=F(I-1)+F(I-2)
100  NEXT I
110  PRINT "THE"K"-TH APPROXIMATION IS"
120  PRINT F(K)/F(K-1)
130  END
```

Notice that the flowchart is basically only a skeleton. Lines 10 and 20, in particular, correspond to no "box" in Figure 2.2.

Now that we have written a program, let's look it over carefully and ask what might go wrong. Think in terms of your mischievous friend who might get his/her hands on the machine. What if he/she INPUTs a hundred jillion for K? Or, maybe just -5? Let's head-off this possibility with an IF . . . THEN instruction:

```
 40  IF K<3 OR K>1001 THEN 130
```

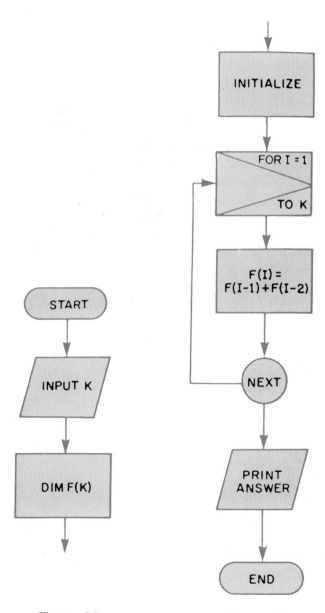

Figure 2.2

Figure 2.3

(Note the new BASIC code word OR and how it is used.) RUN the program, and record the responses for several (increasing) values of K. (Start with some small K's so you can confirm the results by hand as a check that your program is working as planned.)

As a sort of road map, a flowchart is useful not only in writing programs but in finding the mistakes (or "bugs") when your program doesn't work. If there is a logical error, it should show up in the flowchart. In an incidental way, flowcharts promote good programming technique. A cluttered (ugly) flowchart will rarely result in a good program. On a big job, you should always begin with a rough draft of a flowchart. Then look for ways to improve the algorithm(s). Don't be afraid to rewrite your flowcharts several times before starting to work on the actual program. Finally, a flowchart can be used to "document" your program. The idea is that you might want to make changes in a successful program long after you've forgotten the detailed ideas that went into it. The flowchart will help you recall the overall plan.

Speaking of good technique, the program above is very wasteful of memory. (In this case the computer has plenty of memory for the job. But, we might not always be so lucky.) The problem is that we have required the computer not only to compute, but also to remember every Fibonacci number up to and including the K-th. In fact, we have no use for all of them. Consider these lines:

```
 40  A = 1
 50  B = 1
 60  FOR  I = 3  TO  K
 70  C = A + B
 80  A = B
 90  B = C
100  NEXT  I
110  . . .
120  PRINT  B / A
```

In this version, the computer never has to remember more than three things, no matter how big K is. Let's analyze this variation. Lines 40–70 are straightforward. Up to line 80, we have three successive Fibonacci numbers, A, B, and C. We want to produce the next one, namely, B + C. Here is the clever trick. Since we no longer have any use for A, we'll tell the computer to forget it, remembering B instead. The instruction A = B tells the computer to put the current content of memory location B into location A. (Hence, the middle one of our three

Fibonacci numbers in the current step becomes the first one in the next step!) Next (line 90) tell the computer to store the number currently residing in location C in memory location B. (So, what was the third number in our string of three Fibonacci numbers becomes the second number in the next string of three numbers.) Finally, NEXT I takes us back to the computation of (a new) C.

ENTER the new lines 40–100 and 120. RUN this variation, and compare the answers with those you got from the first version.

Exercises (2.4)

1. Figure 2.4 is a flowchart for the program in Exercise 7, Section 2.3. The boxes are labeled with letters. Which lines of the program correspond to each box of the flowchart? (*Hint:* Box A corresponds to no lines of the program, while box B corresponds to line 30. Lines 10 and 20 don't correspond to any box.)

2. Redraw the entire flowchart of Figures 2.2 and 2.3, incorporating the "mischievous friend" program line:

```
40  IF  K<3  OR  K>1001  THEN  130
```

3. Given a positive number, N, can you find the smallest whole number K such that $2^K>N$? Here is a program to do the job. Draw a flowchart to illustrate the program.

```
10  PRINT  "INPUT A POSITIVE NUMBER."
20  INPUT N
40  K=1
50  IF 2↑K>N THEN 80
60  K=K+1
70  GO TO 50
80  PRINT "2↑"K">"N
90  END
```

4. Redraw the flowchart from Exercise 3 to incorporate

```
30  IF  N<1  THEN  90
```

5. Explain what would go wrong if lines 80 and 90 were interchanged in the last program in the text, that is if they become

```
80  B=C
90  A=B
```

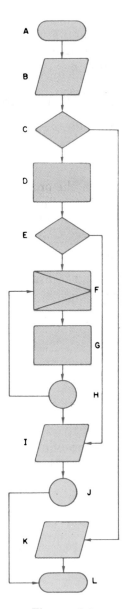

Figure   2.4

6. Generally, it is a tedious procedure to compute the reciprocal of a number with a long decimal expansion. (Consider $1/\pi$, for example.) An interesting property of ϕ is that $1/\phi = \phi - 1$. Using one of the programs in this section, compute ϕ to five decimal places. Call the result x. See how close $1/x$ comes to $x - 1$.

7. Flowcharts needn't be limited to schematic pictures for computer programs. Many algorithms can be illustrated using flowcharts. Draw a "postal" flowchart for a letter mailed from your home to a friend in another city. Include the possibility that the letter may carry insufficient postage and the possibility that the address is incorrect.

8. Suppose you were asked to name a power of 2 that begins with 5. After a moment's thought, you would no doubt come up with 512 $(= 2^9)$. Naming a power of 2 that begins with 6 is even easier: 64 $(= 2^6)$. It is commonly believed that 7 is a lucky number. Can you find a power of 2 that begins with 7? (*Hint:* If 2^n is to begin with 7, it must occur in one of the intervals

$$7 \leqslant 2^n < 8$$
$$70 \leqslant 2^n < 80$$
$$700 \leqslant 2^n < 800$$
$$7000 \leqslant 2^n < 8000$$
$$\cdots$$

Why? How many powers of 2 can occur in one of these intervals? Use an appropriately modified version of the program in Exercise 3 to search for a power of 2 beginning with 7. Finally, try INPUTting 7E6 in line 20 rather than 7000000.)

9. In Exercise 8, you were asked to find a power of 2 that begins with the "lucky" number 7. Think how much harder it will be to find a power of 2 that begins with the "unlucky" number 13. After thinking about it, do it. (Find the smallest power of 2 that begins with 13. *Hint:*

$$13 \leqslant 2^n < 14$$
$$130 \leqslant 2^n < 140$$
$$1300 \leqslant 2^n < 1400$$
$$\cdots)$$

2.5   SORTING

Computers are ideally suited to tedious, repetitive tasks, in the sense that they are willing to perform such tasks provided they are told exactly what to do. That's the rub. The appearance of computers has resulted in a great deal more attention being paid to the structure of algorithms. As you know, an algorithm is a sort of recipe. It is a step-by-step procedure for accomplishing a particular goal.

Consider the following simple illustration. How would you tell a computer to rearrange three numbers so as to put them in increasing order? This, of course, is another of the tasks we don't need a computer for; but that's only because there are a mere three numbers involved. What if there were 3000? Eventually, we may want a computer to rearrange 3000 numbers. There is no magic button to push and no special BASIC code word to accomplish this task. If we want the computer to do the job, we have got to tell it how. And, before we can write a program to tell it how, we need to have an algorithm ready. The real strength of flowcharts lies in their usefulness for devising and describing algorithms.

The first approach that comes to mind (and, therefore, probably not the best approach) is to start by finding the smallest number. Suppose, for the purpose of this discussion, that the numbers are A, B, and C, the obvious thing to do is compare A and B, and then compare the smaller of the two with C. Figure 2.5 is a schematic illustration of this idea.

Even for a simple problem, the algorithm (and flowchart) needn't be simple. Is it obvious that the flowchart in Figure 2.5 illustrates a correct algorithm? If it isn't obvious to you, try the following test. Consider three cases: Suppose A happened to be the smallest. Trace through the flowchart under that assumption and see if it produces the correct response. After you convince yourself that it works in that case, you are ready for case 2, namely, what happens if B is the smallest. Finally, go on to case 3.

Having found the smallest of the three numbers, it would seem to be a simple matter to find the smaller of the remaining two, and, so, finish the problem. But, is it? Which of A, B, and C turned out to be the smallest? We have no way of knowing without having a look at the numbers the symbols A, B, and C represent, and this we can't do! Our problem is to devise the rest of the algorithm without knowing

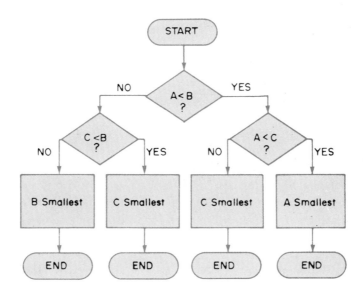

Figure   2.5

which of A, B, or C turned out to be the smallest. (Does this remind you of the old "shell game" in which a pea is hidden beneath one of three walnut shells?)

Have another look at the flowchart. In either of the ways in which C turned out to be the smallest, the complete ordering of the three numbers is clear. On the other hand, if A turned out smallest, then all we can say for sure is that A<B and A<C. There is no information on the relationship between B and C. With this in mind, we can revise our first flowchart and obtain one that completely solves the problem.

The good news about Figure 2.6 is that all six possible orderings are accounted for. The bad news is that there are now six different cases to trace through if you are not convinced that it illustrates a correct algorithm. (But, don't be intimidated! If you're not convinced, try at least some of the six possibilities. What if B<A and A<C. Does the algorithm sort the numbers properly in this case?)

Unfortunately, there is more bad news. The algorithm is only the first step in writing the program.† On the other hand, it is the most

---

†While it may be only the first step, it is usually the most difficult. In any case, it is best to understand sooner, rather than later, that there is no hope of writing the program until the algorithm has been worked out. In fact, more is true. There is no point in even sitting down at the keyboard until you understand the algorithm.

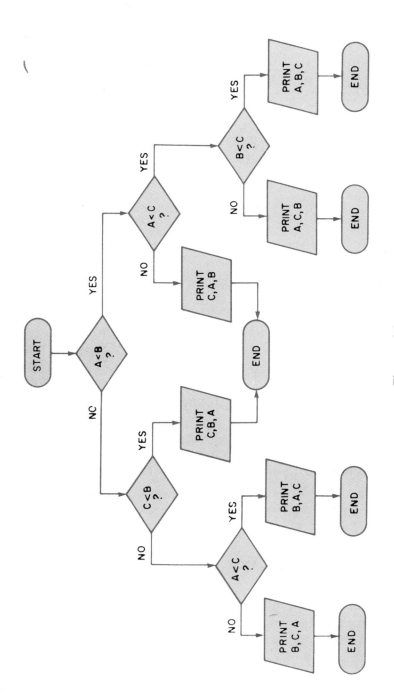

Figure 2.6

important step. Keeping Figure 2.6 before us, it is fairly straightforward to produce the program. Let's begin with

```
10  INPUT  A
20  INPUT  B
30  INPUT  C
```

(If you elect to use the abbreviation INPUT A, B, C in line 10, then leave lines 20 and 30 empty. We'll have use for them in a later revision.) Now we come to the first decision: Is A<B? Or, more particularly, IF A<B THEN what? Then we'll want to go to a line number which is safely out of the way.

```
40  IF  A<B  THEN  500
```

That should do it! Now, line 50 will involve the case that A is not less than B, leading us to the question of whether C is less than B.

```
50  IF  C<B  THEN  400
```

Since we know what happens if C<B, let's take care of that right now (remembering that the order in which the lines are typed in is unimportant).

```
400  PRINT  C,B,A
```

Of course, after 400 is executed (if it is), the computer will proceed to the next numbered line. Let's make it

```
410  END
```

Now, if C is not less than B at line 50, we branch to our next decision: Is A<C?

```
 60  IF  A<C  THEN  300
300  PRINT  B,A,C
310  END
 70  PRINT  B,C,A
 80  END
```

This completes the entire left half of the flowchart (in Figure 2.6). We now come back and pick up the thread in case A<B in line 40. Namely, we proceed to line 500 and compare A with C.

```
500  IF  A<C  THEN  600
510  PRINT  C,A,B
520  END
```

Finally, if A<C,

```
600  IF  B<C  THEN  700
610  PRINT  A,C,B
620  END
700  PRINT  A,B,C
710  END
```

ENTER the program and RUN it, but don't be surprised if it doesn't work the first time. With this many lines, there are lots of opportunities to make mistakes! The computer will catch some of the them and PRINT the SYNTAX ERROR message. Others, it won't catch. (For example, if you type "<" when you mean ">"! Fortunately, you will be able to tell by the response if an error of this type has been made because the numbers will be PRINTed in the wrong order.)

LIST the program, and confirm that the computer has organized the instructions in numerical order, and not in the order in which you typed them in. What happens if two of the numbers are equal? Does the program still work? What if one or more of the numbers is negative?

Let's give the algorithm a real test. Instead of INPUTting three numbers, let's have the computer sort three "random" numbers. It turns out that the computer has a built-in random number generator. ENTER this instruction (in the "interactive" mode, that is, with no line number).

```
PRINT  RND(1)
```

It is expected that the computer responded with a (random) number (strictly) between 0 and 1. If your computer output 1, then try PRINT-ing RND(0). If this works, you will have to replace RND(1) throughout the rest of the text with RND(0). Let's make use of this RaNDom number generating capacity of the machine in our program. Replace lines 10–30 with

```
10  A=RND(1)
20  B=RND(1)
30  C=RND(1)
```

(In spite of appearances, the computer should generate three differ-ent numbers for A, B, and C.) ENTER these new lines, and RUN the revised program several times. You will notice something unsatisfying about the responses. Namely, you won't be able to see that the computer has done anything. Let's fix that.

```
35 PRINT "THE NUMBERS"
36 PRINT A,B,C
37 PRINT "IN INCREASING ORDER ARE:"
```

(Notice how nice it is to have space between lines 30 and 40 for this afterthought.)

One final topic: If you wanted to SAVE† this program, and had the means to do it, you might wonder one day while cleaning out your files what on Earth it was about. We have already talked about the use of flowcharts to "document" programs, so that you can answer just this sort of question. There is another way to document your programs. It involves the REMark statement. In BASIC, any line beginning with the code word REM is ignored by the computer. It is kept as part of the program, but is ignored during program execution. If you are able to save this program, consider documenting it with the following additional lines.

```
800 REM PROGRAM TO SORT 3 NUMBERS
810 REM [YOUR NAME], [THE DATE]
```

Notice that no quotation marks or special "syntax" is needed in REM statements. The computer never reads past the REM.

Exercises   (2.5)

1. Write down an algorithm for finding the largest of three numbers A, B, C.

2. Draw a flowchart to illustrate schematically the algorithm you described in Exercise 1.

3. Write down an algorithm for finding the smallest of four numbers.

4. Draw a flowchart to illustrate your solution to Exercise 3.

5. Here is another algorithm for "sorting" three numbers A, B, C. The idea is to use successive interchanges to put the numbers in increasing order. The first step is to compare A and B. If $A<B$,

---

†SAVE is a BASIC code word to be used if your machine is hooked up to some sort of storage device. Your instructor will be able to tell you if such capabilities are available.

we go to step 2. If not, we interchange the labels on A and B so that the number we now call A is the smaller of the two.† Step 2 involves a comparison of C with the number presently called B. (If the original B was smaller than the original A, then the number presently called B is the number that was originally called A. Otherwise, the number presently called B is the number that was originally called B.) If B<C, we're finished. Otherwise, interchange B and C. Let's call the procedure up to this point the "first pass" through the numbers. It is possible that the three numbers are now in increasing order. Certainly, the largest number is now last. But, it may be that the original C was the smallest number in which case it is now second. What we need to do is compare (and maybe interchange the possibly new) A and B. The flowchart in Figure 2.7 illustrates this second algorithm. Write a program to implement it.

6. If you did Exercises 3 and 4, you don't need to be convinced that a new idea is needed before we'll be ready to sort 3000 numbers (or even 10). One idea is to subscript the numbers. If there are 10 numbers, we should call them $N(1)$, $N(2)$, ..., $N(10)$, not A, B, ..., J. This makes it easy to involve FOR ... NEXT loops in our sorting. Secondly, let's give a name to (i.e., reserve a memory location for) the smallest number in a given set, say S. The flowchart in Figure 2.8 illustrates an algorithm for finding the smallest of the numbers $N(1)$, $N(2)$, ... (We suppose there are K of them.) What we would like to do after we find the smallest number is eliminate it from the sequence. This brings us back to the shell game. Which one was it? Why not just keep track of the location of the number as we keep track of its value? To do this, we need a new name, say L. (This clears up the presence of L in Figure 2.8.) Figure 2.9 illustrates an algorithm for eliminating $N(L)$ from the sequence. Figure 2.10 assembles the pieces into a sorting algorithm. Write a program to implement the flowchart in Figure 2.10. Begin by INPUTting K and generating K random numbers. Incorporate some class into your program with appropriate PRINT statements.

---

†There is a small difficulty involved in interchanging two numbers, say A and B. If we just LET A = B, we lose the original content of memory location A. One way to get around this problem is to store the current content of memory location A in a temporary location, say T. That is, perform the following three steps: T = A, A = B, B = T.

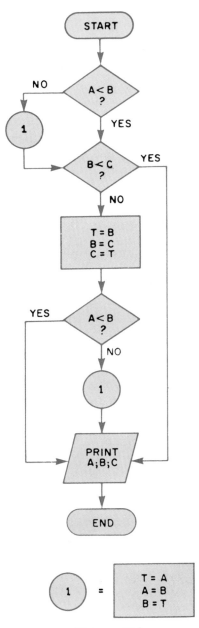

Figure   2.7†

---

†The symbol consisting of a 1 with a circle around it indicates a "subroutine," i.e., a small program that is used (usually more than once) within a larger program. We will have more occasions to use subroutines.

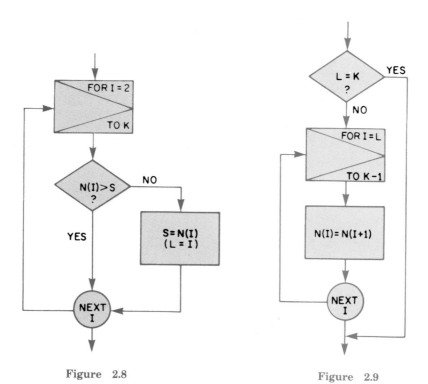

Figure 2.8   Figure 2.9

7. RUN your program from Exercise 6. After you "catch all the bugs," see how long it takes to sort 100 numbers. (RUN the program several times with K = 100 and time each RUN.)

8. Figure 2.11 shows how to extend the method of Exercise 5 to a sorting procedure for K numbers. Write a program to implement it.

9. After you get your program from Exercise 8 working, time several RUNs with K = 100. Compare the computation time of this program with the results you obtained in Exercise 7.

10. The program you wrote in Exercise 6 is probably about twice as fast as the program from Exercise 8. Provided that K = 100 in both programs. Can you suggest why? (*Hint:* As a first step, look at the flowcharts in Figures 2.6 and 2.7. Count the number of steps [arrows between boxes] in the longest path from START to END in each flowchart.)

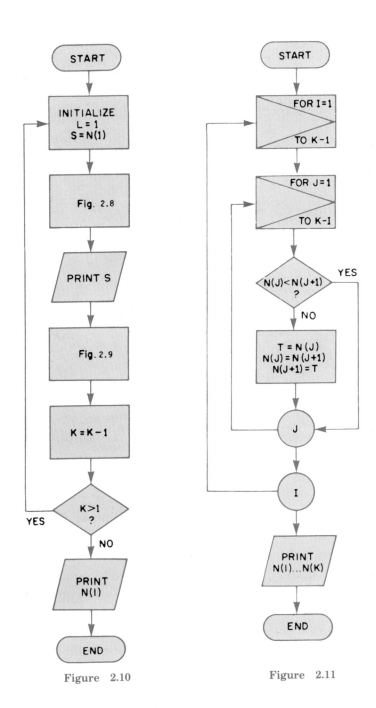

Figure 2.10

Figure 2.11

11. Can you think of any other algorithms for sorting K numbers? If so, draw a flowchart to illustrate your algorithm. (*Hint:* Suppose you had 100 3-by-5 cards each with a different last name on it. How would you alphabetize the cards?)

Chapter 3
PROBABILITY

3.1 SIMULATION

Many games involve an element of chance. We can use the computer's RaNDom number generator to simulate some chance situations. Let's, for example, have the computer simulate the rolling of two dice. What we need to do is figure a way to produce a random whole number oetween 1 and 6 from a random number between 0 and 1. BASIC has a special instruction that makes this easy. It is the INTeger function. Try ENTERing each of the following lines:

```
PRINT  INT(5.934)
PRINT  INT(3.1416)
PRINT  INT(22/7)
PRINT  INT(8)
```

As you see, INT instructs the computer to "return" only the "integer part" of a number. Unfortunately, there is some variation among BASIC dialects concerning what to do with INT of a negative number. Experiment with your machine to determine, for example, if it returns -12 or -13 when you ENTER

```
PRINT  INT(−12.34)
```

To simulate the roll of a single die, it only remains to put INT and RND together. Try RUNning this program

```
10  FOR  I=1  TO  120
20  PRINT  INT(6*RND(1));
30  NEXT  I
40  END
```

(Be warned that computers are very stuffy about parentheses. They insist rather emphatically that every "open" parenthesis be paired with a "close" parenthesis.) Even the most casual glance at the "output" from this program reveals a problem.† It can be corrected by replacing line 20 with the following.

```
20  PRINT  1+INT(6*RND(1));
```

Substitute this new line 20, and RUN the program again.

Now that we have simulated one die, it is a simple matter to simulate two. ENTER and RUN this program:

```
10  FOR  I=1  TO  120
20  PRINT  1+INT(6*RND(1))+1+INT(6*RND(1));
30  NEXT  I
40  END
```

In the context in which we have been using it, a RaNDom number is a number between 0 and 1. In principle, every number between 0 and 1 is equally likely to be chosen. In fact, of course, this is out of the question. A truly random number is unlikely to "terminate" after eight or nine decimal places. A careful look at the concept of "RaNDom" leads to many interesting questions, all of which involve the concept of *probability*.

Consider a situation in which there is a total of D equally likely outcomes. Suppose that N of these are "noteworthy" in some sense. Then the probability that one of the noteworthy events will occur is defined to be the fraction N/D. If, for example, the situation is rolling a fair die, then $D = 6$. There are six equally likely (the defintion of "fair" in this context) outcomes. The probability of rolling a 4, say, is 1/6. Because the only noteworthy event is rolling a 4, $N = 1$. The probability of rolling a 4 *or less* is 2/3. Since rolling any one of 1–4 is "noteworthy," $N = 4$. Since D is still 6, $N/D = 4/6$, or 2/3.

---

† If all 120 numbers are the same, you have even more problems than anticipated! Try replacing RND(1) with RND(0) in line 20.

If we roll two dice, there are 11 possible outcomes; the dice may total any one of the numbers 2–12. But, these 11 outcomes are not equally likely! A glance at Figure 3.1 shows that there are actually 36 equally likely events. (In sorting them all out, it will help to use dice of two different colors, say red and green.) To compute the probability of rolling, say, a 7, we observe that $D = 36$ and $N = 6$. The 6 noteworthy events are: a red 1 and green 6 (abbreviated R1G6), R2G5 (a red 2 and a green 5), R3G4, R4G3, R5G2, and R6G1. Thus, $6/36 = 1/6$ is the probability of rolling 7 with two dice.

We have seen a mathematical definition of probability, but what does it mean in a practical sort of way? To say that the probability is 1/6 of rolling 7 with two dice means, roughly, that if two (fair) dice are rolled a "large number" of times then 7 should turn up about one sixth of the time. If the dice are rolled 120 times, we expect to see 7 about 20 times. If the dice are rolled T times, the expected number of 7's is $T/6$, and this estimate should become better and better, as a percentage of T, the larger T becomes. But, there is no guarantee that there will be any 7's at all!

RUN the program to simulate 120 rolls of two dice again. On a piece of notebook paper, write down a (vertical) column of numbers, 2 through

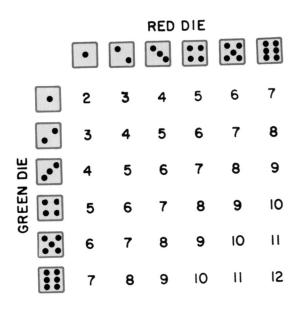

Figure 3.1

12. Read through the output of your RUN and put a "hash mark" (i.e.,
) on your paper beside each number that you come to. When you have
read through the entire list of 120 numbers, you should have 120 hash
marks on your paper. Count and record the number of hash marks
beside each of the numbers 2–12. You have made a "frequency distribution" of the 11 numbers. We can use it to compare the "expected
frequency" of 7 with the actual, or "experimental" frequency.

Exercises (3.1)

1. What is the probability that the name of a randomly chosen month
of the year begins with the letter J?

2. According to an old adage, it is safe to eat shellfish only during
months whose names contain an R. What is the probability that
it is safe to eat shellfish (according to the adage) during a randomly
chosen month of the year?

3. What response would your computer make to PRINT INT(-13.7)?

4. Suppose two fair dice are rolled. What is the probability that their
sum will be
   a. 5? b. 9? c. 12? d. 4? e. 6? f. 1? g.11?

5. Suppose a fair coin is flipped five times and five successive "heads"
are the result. What is the probability that "heads" will occur the
next time the coin is flipped?

6. Write a program to simulate the flip of a fair coin.

7. Write a program to simulate the rolling of five dice.

8. Incorporate your program from Exercise 7 into a FOR ... NEXT
loop to simulate 150 rolls of 5 dice.

9. RUN your program from Exercise 8, and make a "frequency distribution" of the resulting numbers (5–30).

10. Write a program to simulate a roll of one 12-sided die. (A 12-sided
die is made from a "regular dodecahedron," a polyhedron consisting
of 12 pentagonal faces numbered from 1 to 12. See Exercise 16.)

11. Write a program to simulate the sum resulting from a roll of two
12-sided dice. (See Exercise 10.)

12. Have you ever seen a 13-sided die? Probably not. Thirteen may be
an unlucky number, but that isn't the reason there are no 13-sided

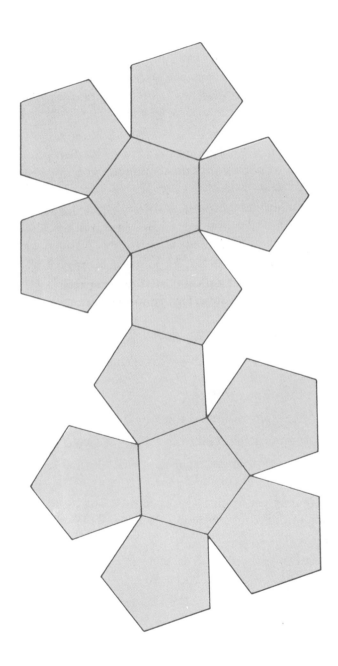

Figure 3.2

dice. (In fact, that is the best reason to produce them if they could be made.) The reason there are no 13-sided dice is that there is no regular polyhedron with 13 faces. But, the computer can still simulate the outcome of rolling a 13-sided die. Write a program to do just that.

13. Write a program to have the computer simulate 1000 rolls of two (6-sided) dice and output the corresponding frequency distribution. (This can be done without any sorting.)

14. If you instruct the computer to PRINT 12/5, you will get a number expressed in decimal form for an answer. Write a program (using INT) to output a quotient and remainder when you INPUT numbers like 12 and 5.

15. Write a program to simulate the rolling of a "loaded" die, one for which a 6 will come up 3/4 of the time and a 1 the remaining 1/4.

16. Trace the picture from Figure 3.2 on a piece of transparent paper. Place the tracing over a piece of stiff paper or thin cardboard. With a ballpoint pen, press down firmly on the "corners" of the drawing. Take away the transparent paper and, with a straightedge and pencil, lightly reproduce the drawing. Finally, cut out the figure and fold along the edges to produce a dodecahedron.

3.2 THE FUNDAMENTAL COUNTING PRINCIPLE

It turns out that the definition of probability is deceptively simple. It is often very difficult to compute probabilities correctly. The biggest difficulty encountered by beginners centers around the "equally likely" restriction. When, for example, a meteorologist predicts a 10% chance of rain for tomorrow, he/she is estimating that the probability of rain is 1/10. But how can this be? There are only two possible outcomes. Either it will rain, or it won't. The single noteworthy outcome is rain. Why, then, is the probability of rain not equal to 1/2? The answer, of course, is that the two outcomes are *not equally likely*. What the meteorologist is saying is that rain is much less likely than no rain. Indeed, the odds against rain are 9 to 1.[†]

The second major difficulty with computing probability involves actually determining the numbers N and D. The problem is counting! Everyone believes they know how to count, and indeed they do, provided the items to be counted are nicely arranged, preferably in a straight line. But what about determining the number of jellybeans that will fit in a school bus?

There is a powerful principle that is very useful in many counting problems. It is known as the *Fundamental Counting Principle*. Think of it this way. Suppose there are K decisions to be made. If there are $C(1)$ choices for the first decision, $C(2)$ choices for the second decision, ..., and $C(K)$ choices for the Kth decision, then the total number of different ways the sequence of decisions can be made is the product $C(1)$ times $C(2)$ times ... times $C(K)$. Take the case of rolling two dice— a red one and a green one. We may think of this as a sequence of two decisions. The first decision (made by "Lady Luck") is what number should come up on the red die. The second decision is what number should come up on the green die. Thus, $K = 2$, $C(1) = 6$, and $C(2) = 6$. The Fundamental Counting Principle asserts that the total number of different ways the sequence of two decisions can be made, that is, the total number of possible outcomes when two dice are thrown, is 6 times $6 = 36$. This answer confirms the brute-force enumeration exhibited in Figure 3.1 in the previous section. If the dice are "fair," then each of these 36 possible outcomes is equally likely.

Let's try another example. Suppose the four letters in the word "LUCK" are written on four marbles. Imagine the marbles put in a brown paper bag and rolled around. Suppose they are then taken out, one at a time, at random. If we make up a "word" by writing down each letter in the order in which it is drawn from the bag, how many different 4-letter "words" could result? The answer is a simple application of the Fundamental Counting Principle: The first "decision" is which marble to draw first. Because there are four marbles, the number of choices for the first decision is four. The second decision is which marble to draw second. At this point, one marble has already been taken from the bag,

---

[†] To convert from "odds" to probability, let the denominator, D, be the sum of the two numbers making up the odds. The numerator, N, is then one of the two numbers making up the odds. If the odds are 5 to 3 that the Raiders will win the Super Bowl, then the estimated probability (by the odds-makers) that the Raiders will win is 5/8.

leaving three marbles to be drawn. Thus, $C(2)$, the number of choices for the second decison, is three. Similarly, $C(3) = 2$ and $C(4) = 1$. The total number of ways to make the sequence of decisions (the total number of "words," including nonsense words that can be made up from the letters L, U, C, and K) is $4*3*2*1 = 4!$[†] Here is a list of all 24 possibilities. Look it over carefully and convince yourself that every possible arrangement of the four letters is present and that none is listed twice.

LUCK	ULCK	CLUK	KLUC
LUKC	ULKC	CLKU	KLCU
LCUK	UCLK	CULK	KULC
LCKU	UCKL	CUKL	KUCL
LKUC	UKLC	CKLU	KCLU
LKCU	UKCL	CKUL	KCUL

(While some of these "words" may be "call letters" for radio stations in your area, few of them have any other meaning.)

If you have extra time, here is a game you can ENTER and play:

```
 10  REM GUESS THE NUMBER GAME
 20  N=1+INT(100*RND(1))
 30  PRINT "I HAVE A NUMBER BETWEEN 1 AND"
 40  PRINT "100. TRY TO GUESS WHAT IT IS."
 50  FOR I=1 TO 8
 60  INPUT K
 70  IF K=N THEN 300
 80  IF K<N THEN 150
 90  PRINT "GUESS TOO BIG"
100  NEXT I
110  GO TO 200
150  PRINT "GUESS TOO SMALL"
160  NEXT I
200  PRINT "BETTER LUCK NEXT TIME."
210  END
300  PRINT "RIGHT! YOU GUESSED MY NUMBER"
310  PRINT "IN ONLY" ; I ; "TRIES."
320  END
```

[†]See Exercise 13, Section 2.3, for an explanation of the "factorial" notation.

Exercises   (3.2)

1. The (written) Hawaiian language has only 12 letters, the vowels a, e, i, o, u, and the consonants h, k, l, m, n, p, and w. Using the Hawaiian alphabet, how many
   a. 3-letter "words" are possible?
   b. 3-letter "words" are possible if the middle letter is a vowel?
   c. 3-letter "words" are possible if the middle letter is a vowel and the other two letters are consonants?

2. Using the Hawaiian alphabet (see Exercise 1), what is the probability that a random 3-letter "word" will begin and end with consonants and have a vowel in the middle?

3. Suppose you are handed a pair of unusual (6-sided) dice. Suppose the red one contains only even numbers, two 2's, two 4's, and two 6's. If the green one has only odd numbers, two 1's, two 3's, and two 5's, write out a table (similar to the one in Figure 3.1) showing all possible (equally likely) outcomes if the dice are rolled.

4. Assume each die in Exercise 3 is "fair" in the sense that any one of the six faces is equally likely to come up when the die is rolled. What is the probability of rolling a 7 with these dice? What is the probability of rolling a 10 with the dice?

5. Suppose you are shown a pair of nonstandard (6-sided) dice, one of which is numbered with the following six numbers: 1, 3, 4, 5, 6, and 8. If the six faces of the other die are numbered with a 1, two 2's, two 3's, and a 4, write out a table (similar to the one in Figure 3.1 and Exercise 3) showing all 36 equally likely outcomes.

6. Compute the probability of rolling each of the sums 2–12 with the dice in Exercise 5, and show that the results are identical with those obtained by rolling two standard dice.

7. Draw a flowchart to illustrate the program for the "GUESS THE NUMBER GAME."

8. Suppose the letters in the word "COMPUTER" are written on blank 3-by-5 cards and rearranged randomly to make up an 8-letter "word." Compute the number of possible words, including nonsense words, that can be made up in this way.

9. Use the Fundamental Counting Principle to compute the total number of (equally likely) outcomes when five (ordinary, 6-sided) dice are rolled.

10. What is the probability of rolling five 6's if five fair dice are rolled? (What do you think is the probability of rolling five 6's in a row using a single die? *Hint:* See Exercise 9.)

11. What is the probability that all the dice will come up the same if five 6-sided dice are rolled?

12. Use the Fundamental Counting Principle to show that in a family with four children, there are 16 possible arrangements for the sexes of the children.

13. In a family with four children, what is the probability that half the children are girls and half are boys? (Don't just guess! Make use of Exercise 12, and write down all the equally likely possibilities. Use a code like BGBB to represent the possibility that the second oldest child is a girl while the rest are boys.)

14. Write a program to simulate 1600 4-child families and compute the fraction of them that are half boys and half girls. (*Hint:* Use a numerical code, 1 for a girl, and 2 for a boy. Thus, 2122 represents a family in which the second oldest child is a girl and the other three are boys. Convince yourself that the numbers in such a code will total 6 if, and only if, two of the children are girls and two are boys.)

15. It will be convenient to define $0! = 1$. (Note that this is a separate definition, unique to 0.) Show that

$$40,585 = 4! + 0! + 5! + 8! + 5!$$

Does this kind of decomposition work for all numbers?

16. Some California license plates begin with a (single, numeric) digit followed by three letters followed by three more digits. How many different license plates of this type are possible?

3.3 REARRANGEMENTS

In the last section, we found that $4! = 24$ different "words" could be made by rearranging the letters in the word LUCK. What about re-

arranging the letters in the word LOOK? To simplify the situation, let's
first answer the question when the two O's are distinguished; for example, let's do the problem for the word LOok. In this case, the solution
is the same as for LUCK. There are four choices for the first letter
(namely, L, O, o, and K). Once a letter has been chosen to be first,
there are three (remaining) choices for the second letter, and so on.
Here are the 24 possibilities:

LOoK	OLoK	oLOK	KLOo
LOKo	OLKo	oLKO	KLoO
LoOK	OoLK	oOLK	KOLo
LoKO	OoKL	oOKL	KOoL
LKOo	OKLo	oKLO	KoLO
LKoO	OKoL	oKOL	KoOL

Now, the question is, "How many *different* 4-letter words can be
made by rearranging the letters in the word LOOK?" Notice that if
we do not distinguish big O from little o in our list, every word in the
list occurs exactly twice! Thus, the number of different words is 24/2
= 12. (This is a little like determining the number of people in the
room by counting the number of eyes and dividing by 2. This is a
laughable way to count people, but we will see that it is a serious
method in other contexts.)

Suppose we play the same game with the word LULL. How many
different words can be made up by rearranging the letters in the word
LULL? Is the answer 24/3 = 8? We can easily see that this is wrong.
As soon as we decide where to put the U, our word is completely
determined. Since there are only four different spots for the U, only
four different words can be made from the letters in LULL! What is
wrong with 24/3? To answer this question, we have to ask another:
Why divide by 3? Presumably, because if we distinguish the 3 L's and
write out all 24 possibilities, each different word would occur three
times. Evidently, this is wrong; and a little more thinking shows why.
Let's ask how many times LULL itself would appear in the list of 24
words. Use the Fundamental Counting Principle: The first L can occur
in any one of three places. The second L can occupy any one of the two
remaining places, by which time the location of the third L is determined. Thus, there are 3*2*1 = 3! = 6 ways of arranging the 3 L's
into the framework _U_ _ to obtain LULL. Since each of these will
occur in the list of 24 exactly once, LULL occurs six times. A similar

argument can be made for ULLL, and each of the other possibilities. So, in our list of 24 words, each word occurs a total of six times. The number of different words occurring, therefore, is 24/6 = 4.

Let's summarize. There are 4! different four letter "words" that can be made by rearranging the letters in the word LUCK. There are 4!/ 2 ways to rearrange the letters in LOOK, and 4!/3! rearrangements of LULL. This suggests another question. If we divide by 3-factorial for LULL, why didn't we divide by 2-factorial for LOOK? The answer, of course, is that we did: 2! = 2.

Are you ready for MISSISSIPPI? How many 11-letter words can be made by rearranging the letters in the word MISSISSIPPI? If all the letters were different, the answer would be 11!. To make up for the fact that there are 4 I's, we need to divide by 4!. To make up for the fact that there are 4 S's, we need to divide the result (namely, 11!/4!) by (another) 4!. Finally, to make up for the 2 P's, we need to divide that result by 2!. So, the number of different words that can be made up by rearranging the letters in the word MISSISSIPPI is ((11!/4!)/4!)/ 2! = 11!/(4!*4!*2!), or 34,650.

Let's write a program to calculate numbers of the form $A!/$ $(B!*C!*...*D!)$. The first step is a program to compute the factorial of a number. (See Exercise 13, Section 2.3.) ENTER this program, and confirm that it produces the correct answer by comparing it with a paper-and-pencil calculation of 3!, 4!, ..., 6!. (Note that 4! = 4*(3!), etc.)

```
 10  PRINT "I COMPUTE N!"
 20  INPUT "N=";N
200  F=1
210  FOR I=2 TO N
220  F=F*I
230  NEXT I
240  PRINT "N! =" ; F
```

We want to use this program as a "subroutine," that is, as a program within a program. This can be accomplished by means of the BASIC code words GOSUB and RETURN. The command GOSUB, like GO TO, instructs the computer to execute the program commands in some order other than sequential (numerical) order. GOSUB XXX directs the computer to line number XXX. Program execution continues from line XXX until the RETURN command is encountered. At that point, program execution reverts to the line immediately following the GO-

SUB XXX instruction. The advantage of GOSUB . . . RETURN is that the subroutine can be used several different times (at several different places) in the program, and execution will always revert to the correct place. Amend the program by ENTERing these lines. (Don't type NEW. Just add/change the lines below. Lines 200–230 will remain unchanged.)

```
 10  PRINT"I COMPUTE A!/(B(1)!*...*B(K)!)"
 20  INPUT " A=";A
 30  INPUT " K=";K
 40  DIM B(K)
 50  FOR I=1 TO K
 60  PRINT "INPUT B(";I;")."
 70  INPUT B(I)
 80  NEXT I
 90  N=A
100  GOSUB 200
110  A=F
120  D=1
130  FOR J=1 TO K
140  N=B(J)
150  GOSUB 200
160  D=D*F
170  NEXT J
180  PRINT "THE ANSWER IS"; A/D
190  END

240  RETURN
```

Try the program on several examples in which you are able to confirm the answer by means of a hand calculation. In particular, check to see that 11!/(4!*4!*2!) = 34,650. (Of course, you can save a lot of time and effort in your hand calculations by cancelling common factors; for example,

$$11!/(4!*4!*2!) = (11!/4!)/(4!*2)$$

$$= (11*10*9*8*7*6*5)/(4*3*2*2)$$

$$= (11*10*9*8*7*6*5)/(8*6)$$

$$= 11*10*9*7*5.)$$

Note: The program assumes that **A**, **K**, and all the B(I)'s are greater than 1. Since 1! = 1, the reader has the option of incorporating the 1-factorials in his/her head or of amending the program by adding

```
205  IF  N = 1  THEN  240
```

Exercises *(3.3)*

1 Suppose the pot of gold at the end of a rainbow contains 15 $20 gold pieces, 45 old Roman coins worth $100 each, 12 gold sovereigns, and 3 nickels. If a single coin were to fall into your hand just as the rainbow faded away, what is the probability it would be a nickel?

2. What is the largest possible probability a noteworthy event can have? The smallest?

3. How many outcomes are possible if three dice are rolled, a 6-sided, an 8-sided, and a 12-sided die?

4. How many different words can be made by rearranging the letters in
   a. WORDS   b. LETTERS   c. REARRANGING
   d. TENNESSEE

5. Explain why line 90 is needed in the program.

6. If you input A = 5 in line 20 of the program, what does A become in line 110?

7. Explain why it is necesary to use J (or, at least a variable different from I) in line 130 of the program. Why is it okay to use I in line 210?

8. Why is line 120 needed in the program?

9. What happens if you input A = 100, K = 1, and B(1) = 1 while RUNning the program?

10. Suppose you learn that in a particular two-child family, one (at least) of the children is a boy. What is the probability that the other is a boy? (Don't just guess. Work it out as if your life depended on getting the right answer!)

11. Suppose you get a job in San Francisco for the summer. Suppose further that you have a favorite aunt who happens to own an

apartment at the corner of Post and Powell in the city. If you stay
with your aunt, and walk to work at the corner of Clay and Larkin,
you will have a choice of many routes. (See Figure 3.3.) Assuming
that you always walk either north or west at each intersection (so
that you always walk exactly 12 blocks), how many different routes
are there between work and your aunt's apartment? (*Hint:* Consider using a code to identify each route. The code for the route in
Figure 3.4 might be NWNNWWWWWNNNW. Observe that every
12-block route from Post and Powell to Clay and Larkin involves
exactly 6 "north-blocks" and 6 "west-blocks.")

12. Suppose the King and Queen of hearts are shuffled together with
the King and Queen of spades. If two cards are drawn at random
from these four, what is the probability that they will be of the
same suit?

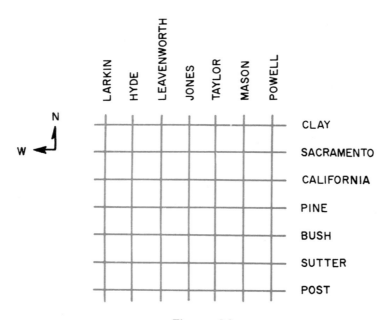

Figure   3.3

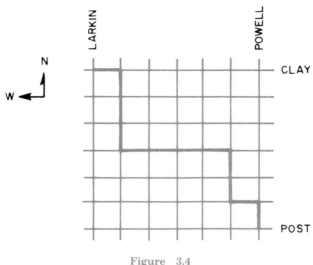

Figure   3.4

3.4   BINOMIAL COEFFICIENTS

Probabilities involving coin flips or sexes of children in a family come under the category of "binomial probability." Consider 12 flips of a fair coin. What is the probability that half the flips are heads and half are tails? By the Fundamental Counting Principle, there are 12 "decisions" to be made, each decision involving two choices. Thus, the total number of (equally likely) possibilities is $2^{12} = 4096$. That's the easy part. Of these 4096 outcomes, how many are noteworthy? It will help to use a code. Let HTTTHHTHTHHH represent the outcome that the first flip comes up heads, the next three tails, followed by two heads, etc. Then there is a natural correspondence between the 4096 coin-toss outcomes and the 4096 12-letter words that can be made using the two letters H and T. There is more! The number of noteworthy outcomes is exactly the number of 12-letter words that can be made by arranging 6 H's and 6 T's, precisely the problem we solved in the last section. The desired number is $12!/(6!*6!) = 924$. Thus, the probability of flipping half heads and half tails is $924/4096 = 0.22558...$, less than 1/4.

Is mathematics real? Does it apply in the real world, or is it just a hard, esoteric subject that one must endure along the road to full intellectual maturity? Don't *believe* it just because it is printed here or because someone with lots of credentials says it. Be skeptical. Don't be satisfied until you have convinced yourself that it is really true. Really! Take the last paragraph, for example. Why not flip a coin 12 times and see what happens? The problem is that doing such an experiment just once won't tell you anything. You might flip 6 heads and 6 tails and you might not. So what? The mathematics merely says that if you perform the experiment a "large number" of times, you will flip 6 heads and 6 tails less than one-fourth of the time. Who wants to flip a coin 12,000 times? Why not have the computer simulate it? ENTER and RUN the following program. (If you get tired of watching before the program ENDs, or if you become convinced that the probability really is .2255, use the program interrupt procedure to stop the program execution.)

```
 10  FOR I=1 TO 1000
 20  H=0
 30  T=0
 40  FOR J=1 TO 12
 50  X=INT(2*RND(1))
 60  IF X=0 THEN 100
 70  PRINT "H";
 80  H=H+1
 90  GO TO 120
100  PRINT "T";
110  T=T+1
120  NEXT J
130  IF H=T THEN PRINT "***";
140  IF H=T THEN N=N+1
150  PRINT N/I
160  NEXT I
170  END
```

In most dialects of BASIC, a program of the type LISTed above can be shortened by a technique called "crunching." The idea is to put more than one instruction on a program line. This is done by separating the instructions by colons. After you get the program RUNning, try this variation. Replace line 20 with

```
 20  H=0 : T=0
```

and eliminate line 30 altogether. (Recall that you can eliminate a line by ENTERing just the line number. If you ENTER

30

then line 30 will be eliminated. Try it, and verify the result by LISTing the revised program.) RUN this variation, and see if the version of BASIC on your machine supports crunching. If the revised program RUNs successfully, try a further revision by eliminating line 140 and replacing line 130 with

130 IF H=T THEN PRINT "***";: N=N+1

Some care has to be taken with crunching lines involving IF ... THEN statements. In the new line 130, if H does not equal T, the computer will not take notice of the rest of the line. The computer executes the rest of the line, including any multiple instructions, only if H = T.

Let's return to the mathematics of 12 coin flips. How different would the computation be if we wanted the probability that a 12-child family consists of 6 boys and 6 girls? If we assume the stork is "fair," then the computation is the same except for superficialities (like using G and B instead of H and T). In a sense, mathematics consists of just those things that remain after such superficialities are stripped away.

What is the probability that 12 flips of a fair coin will result in 5 heads and 7 tails (or 5 tails and 7 heads)? As in the case of 6 heads and 6 tails, we make use of the H-T code (to convert the problem into one involving numbers of words; that is, how many 12-letter words can be made by arranging 5 H's and 7 T's? Just $12!/(5!*7!) = 792$. Thus, the probability of flipping 5 heads and 7 tails (or of having 5 girls and 7 boys in a 12-child family) is $792/4096 = .19335...$, or a little less than 1/5.

Numbers of the form $n!/(r!*(n-r)!)$ occur so often in counting problems that they are given a special name and (more than one) special notation. In deference to the typesetter's union, we will use the notation nCr for this "binomial coefficient," even though $\binom{n}{r}$ is more common. Thus,

$$nCr = \frac{n!}{r!*(n-r)!} \qquad (3.1)$$

It would be well to remember nCr in two ways. One is an "algebraic" way, namely, the formula given above for calculating it. The other way is a "combinatorial" way. It is the number of n-letter words that can be made using two letters in which the first letter is used r times and the second letter the remaining number of times. Some facts are more easily comprehended if you think algebraically, and some more easily understood if you think combinatorially. For example, is it obvious from Equation (3.1), i.e., the algebraic viewpoint, that

$$nC0 + nC1 + nC2 + \cdots + nCn = 2^n? \qquad (3.2)$$

(To make things come out right in expressions like this one, we define 0! (0-factorial) to be 1.) If you're good at manipulating factorials, you might be able to grind out an algebraic proof of this observation. But, the combinatorial proof is much easier. The equation asserts the following: The total number of possible n-letter "words" that can be made from a 2-letter alphabet, for example, A and B, is equal to the sum of the number of words that involve no B's (i.e., $nC0 = 1$), plus the number involving just one B (i.e., $nC1 = n$), plus the number with $(n-2)$ A's and 2 B's (i.e., $nC2$), plus ... plus nCn (the number with n B's). Every word has to have some number of B's and the rest A's. Since we have already determined that the total number of such words is 2^n, the proof is finished.

Exercises (3.4)

1. Compute nCr if
   a. $n=4$ and $r=2$ b. $n=5$ and $r=3$ c. $n=5$ and $r=2$
   d. $n=6$ and $r=2$ e. $n=6$ and $r=3$ f. $n=6$ and $r=4$

2. Confirm Equation (3.2) when $n = 5$, i.e., compute $5C0, 5C1, \ldots,$ and $5C5$. Sum them, and confirm that the answer is 2^5. (*Note:* 0! = 1.)

3. The "call letters" of radio stations in the United States typically begin with K or W. How many different 4-letter "call letters" are possible?

4. Prove that $nCr = nC(n-r)$. Give both an algebraic proof and a combinatorial proof.

5. Confirm Equation (3.2) when $n = 8$, i.e., compute $8C0, 8C1, \ldots,$ $8C8$, sum them, and confirm that the answer is 2^8. (Remember, 0! = 1.)

6. In an 8-child family, compute the probabilty that the number of girls is

   a. 0   b. 1   c. 2   d. 3   e. 4   f. 5   g. 6

   h. 7   i. 8

   (*Hint:* See Exercise 5.)

7. While the probability that an 8-child family consists of 4 girls and 4 boys is not 1/2, from among all the possible sex distributions, 4 girls and 4 boys is the most likely. Verify this statement. (*Hint:* See Exercise 6.)

8. Explain why the use of the letter H in line 70 of the program does not conflict with its (different) meaning in lines 20, 80, and 130–140.

9. If the following program were RUN, what would be its output? (This is a problem for you, not for the computer.)

```
 10 PRINT "S";
 20 GOSUB 170
 30 GO TO 60
 40 PRINT "HELP ME"
 50 END
 60 GOSUB 100
 70 PRINT "D M";
 80 GO TO 150
 90 PRINT "?!"
100 PRINT "N";
110 RETURN
120 GOSUB 100: GOSUB 170
130 PRINT "Y"
140 PRINT "PLEASE!": END
150 PRINT "O";
160 GO TO 120
170 ? "E";
180 RETURN
190 GO TO 10
```

10. Using a standard 52-card deck, how many different 5-card Poker hands are possible?

11. Prove that $(r/n)*(nCr) = (n-1)C(r-1)$.

3.5 ORDERED SELECTIONS

Suppose your Aunt Vera decides it's time to move from her large house into a small apartment. In going through her things, she comes across her old record collection of Rock music from the 1950s. She decides to give the records away to you and your cousins and invites everyone over to choose his/her favorite albums. Looking over the 112 different records, you find that you have a definite first, second, third, fourth, and fifth choice. Among the remainder, there are some you wouldn't want to be seen with, some you could take or leave, and a few you would be glad to have. But, none of the others comes close to your first five choices; and, if worst comes to worst, you have a definite ranking of favorites among those five.

It turns out that each of your cousins feels the same way. Each has five choices that he/she prefers to any of the others, and each has a definite preference ranking among the five choices. If you and each of your cousins were to write down his/her list of five, in rank order, then a number of possibilities might occur. When two of your cousins compare lists, it could happen that they don't overlap at all, that they overlap in part, that they contain the same five albums but in a different order, or that the two lists are identical. For the moment, let's say that two lists are the same only if they contain the same albums, in the same rank order. In this case, how many different lists are possible? The answer, of course, is an easy application of the Fundamental Counting Principle. As there are 112 possibilities for the first album on the list, 111 possibilities for the album occupying the second preference spot, . . . , and 108 remaining possibilities for the fifth choice spot, the answer is 112*111*110*109*108.

In general, if we want to make a selection of r things from n things in which the order is important (that is, if we have a preference ranking), the number of different ways of doing it is

$$n*(n-1)*(n-2)*\cdots*(n-r+1) \qquad (3.3)$$

(Why the "$+1$"? Try starting from the other end. If we want the top r numbers, they will be $(n-r+1)$, $(n-r+2)$, $(n-r+3)$, . . . , $(n-r+r)$ $= n$.) This kind of selecting occurs so often that mathematicians have devised a special code (or shorthand) for the expression in (3.3). It is

$nPr.$[†] For example, the number of five-album preference lists from Aunt Vera's record collection is 112P5. (Notice that $112 - 5 + 1 = 108$.)

A closely related counting problem is the one of selecting r things from n things in which the order *doesn't* matter. This would be the case, for example, if five of the albums were tied for first place in your preference ranking, and all the remaining 107 albums were less desirable. If order doesn't matter, (if, for example, we consider two lists to be the same if they contain the same five albums, regardless of order) then how many different lists are possible? Of course, 112P5 is too large because it counts each (unordered) list more than once. Well, how many different ways are there to make an ordered list involving the same five albums? Any one of the five could be first; any one of the four remaining albums second. The answer is 5! (which, by the way, is 5P5). So, to count the number of lists in which order doesn't matter, we might first count the number in which it does (i.e., 112P5), and divide by 5! to make up for the fact that each (unordered) list has been counted exactly 5! times. The number of unordered lists is

$$112P5/5! \ = \ 112*111*110*109*108/5! \ = \ 112C5.$$

More generally,

$$nCr \ = \ nPr/r!$$

is the number of ways of choosing r things from n in which order doesn't matter. The code "nCr" is read "n-choose-r" or "binomial coefficient n-over-r."

Of course, the "nCr" from this discussion is the same as the "nCr" of the previous section. It is easy to give an algebraic proof that this is so. (See Exercise 1.) It is probably worth the trouble to give a combinatorial argument as well.

Let's count the number of unordered five-album lists in another way. Imagine the 112 albums lined up in a row. Here is how your (somewhat eccentric) Aunt would like you to communicate your five first-choices to her: Walk down the row of albums and, with a piece of chalk, write "Y" beneath the five albums you most want, and "N" below the other 107 albums. Being from the "old school" she is having you do things the way she learned them! She is having you write down a 112-letter "word" consisting of 5 Y's and 107 N's. How many such 112-letter words

---

[†] Naturally, this secret code is revealed to you in the strictest confidence!

are there? Exactly one for each possible unordered list of five albums. Thus, the number of such albums (as determined in the previous section) is 112C5.

The binomial coefficients nCr occur in many contexts, some of which are not (obviously, at least) connected with counting. One of the most famous of these is the so-called Binomial Theorem. Consider these:

$$(x+y) \;=\; 1{*}x \;+\; 1{*}y$$

$$(x+y)^2 \;=\; 1{*}x^2 \;+\; 2{*}xy \;+\; 1{*}y^2$$

$$(x+y)^3 \;=\; 1{*}x^3 \;+\; 3{*}x^2y \;+\; 3{*}xy^2 \;+\; 1{*}y^3$$

$$(x+y)^4 \;=\; 1{*}x^4 \;+\; 4{*}x^3y \;+\; 6{*}x^2y^2 \;+\; 4{*}xy^3 \;+\; 1{*}y^4$$

$$\cdots$$

$$(x+y)^n \;=\; nC0{*}x^n \;+\; nC1{*}x^{n-1}y \;+\; nC2{*}x^{n-2}y^2$$
$$+\; nC3{*}x^{n-3}y^3 \;+\; \cdots \;+\; nCn{*}y^n \qquad (3.4)$$

Equation (3.4) is called the Binomial Theorem. It expresses the fact that the coefficient of $x^{n-r}{*}y^r$ in the "expansion" of $(x+y)^n$ is the binomial coefficient nCr. In fact, it is not hard to see that this is so. It is a consequence of the distributive rule that

$$(x+y){*}(x+y){*}\ldots{*}(x+y) \qquad (n\text{-times})$$

can be obtained as follows: Choose one of the terms, x or y, from the first bracket, one from the second, one from the third, and so on, finally choosing one from the n-th bracket. Once you have made these n selections, multiply them all together. If, for example, you chose x from the first bracket, x from the second, y from the third, and so on, finally choosing x from the last bracket, then after multiplying, you would have $xxy\ldots x$, a "word" of length n in the two letters x and y. Now, the product $(x+y)^n$ is obtained by making all possible (2^n) such selections and adding up all the resulting words. For example, if $n=3$, the $2^3 =$ 8 "words" are xxx, xxy, xyx, xyy, yxx, yxy, yyx, and yyy. But, since multiplication is commutative, it is usual to group, e.g., xxy, xyx, and yxx together as $3{*}x^2y$. In the general case, we would group together all words involving the same number of x's and y's. So, the coefficient of $x^{n-r}y$ in the expansion $(x+y)^n$ is equal to the number of ways a word with $(n-r)$ x's and r y's can arise in the above selection (one from each bracket) process. But any n-letter word with $(n-r)$ x's and r y's will arise exactly once in that process. Therefore (finally!), the coefficient of $x^{n-r}y^r$ must be equal to the total number of n-letter words that can be made by arranging $(n-r)$ x's and r y's. We know this number to be nCr. Q.E.D.[†]

One amusing consequence of the Binomial Theorem is another proof that $nC0 + nC1 + nC2 + \cdots + nCn = 2^n$. Just take $x=y=1$ in Equation (3.4).

Exercises (3.5)

1. Explain, algebraically, why $nCr = nPr/r!$.

2. Compute ea h of the following:
   a. 7P3   b. 5P1   c. 10P2   d. 11P3   e. 101P3

3. Compute each of the following:
   a. $7C3$ b. $5C1$ c. $10C2$ d. $11C3$ e. $101C3$

4. Compute the coefficient of x^6y^6 in $(x+y)^{12}$. Explain what this has to do with 12-child families in which half the children are girls and half are boys. Explain what it has to do with Exercise 11, Section 3.3.

5. Compute the coefficient of x^3y^{101} in $(x+y)^{104}$.

6. Prove that $nC0 - nC1 + nC2 - nC3 + \cdots \# nCn = 0$, where "#" stands for "+" if n is even and "−" if n is odd.

7. Prove that $nCr = [n/(n-r)]*[(n-1)Cr]$.

8. Confirm the identity in Exercise 7 for the following values of n and r:
   a. $n = 5, r = 3$ b. $n = 6, r = 3$ c. $n = 6, r = 4$.

9. Prove that $(nCr)*(rCk) = (nCk)*[(n-k)C(r-k)]$.

10. Confirm the identity in Exercise 9 for the following values of n, r, and k:
    a. $n = 6, r = 4, k = 2$.   b. $n = 8, r = 5, k = 3$
    c. $n = 9, r = 6, k = 2$.   d. $n = 9, r = 6, k = 3$.

11. Write a program to INPUT n and r, and to output nPr.

[†]"Q.E.D." stands for the Latin, *quod erat demonstrandum*, meaning "which was to be shown."

3.6   PASCAL'S TRIANGLE

As we have seen in the last two sections, the binomial coefficients, nCr, are useful. It would be good to know more about them. At the moment, they are probably still a bit mysterious, maybe even threatening. Try to look upon them as useful allies, potential good friends. It is worth some trouble to become better acquainted with them. Let's do it.

So far, we have two distinct ways of thinking about them. The "algebraic" definition is

$$nCr = \frac{n!}{r!(n-r)!} \tag{3.5}$$

The "combinatorial" definition that arose from considering Aunt Vera's record collection is that nCr is *the number of ways of choosing* r *things from* n *in which order doesn't matter.* Using either of these interpretations, one can deduce that

$$nC0 = 1 \text{ and } nCn = 1. \tag{3.6}$$

Another interesting identity is this:

$$nCr = (n-1)Cr + (n-1)C(r-1). \tag{3.7}$$

In Exercise 3, you will be asked to give an algebraic proof of this identity. Here is a combinatorial proof: We look at nCr as the number of ways to choose r things from a set of n things (in which the order of choosing doesn't matter). Consider the n-th item. If we don't include it, then all r choices must come from the first $(n-1)$ elements. There are $(n-1)Cr$ selections of this type. If, on the other hand, the n-th item is included, then the remaining $(r-1)$ choices must be made from the remaining $(n-1)$ things; there are $(n-1)C(r-1)$ ways to make the selection in this case. Because the last element is either among those selected or it isn't, we have exhausted all possibilities, that is, Equation (3.7) is established.

Equation (3.7) is a so-called *recursion* relation. If we know all about choosing from $(n-1)$-element sets, then we can deduce information about choosing from n-element sets. This identity is the basis of what has come to be known as *Pascal's Triangle*. It is a way to generate and display the binomial coefficients. Consider the array

$1C0$	$1C1$					
$2C0$	$2C1$	$2C2$				
$3C0$	$3C1$	$3C2$	$3C3$			
$4C0$	$4C1$	$4C2$	$4C3$	$4C4$		
$5C0$	$5C1$	$5C2$	$5C3$	$5C4$	$5C5$	
$6C0$	$6C1$	$6C2$	$6C3$	$6C4$	$6C5$	$6C6$

Table 3.1

As an alternative to Equation (3.5) we intend to fill in this table a row at a time. First, however, make use of Equation (3.6) to obtain:

1	1					
1	$2C1$	1				
1	$3C1$	$3C2$	1			
1	$4C1$	$4C2$	$4C3$	1		
1	$5C1$	$5C2$	$5C3$	$5C4$	1	
1	$6C1$	$6C2$	$6C3$	$6C4$	$6C5$	1

Table 3.2

Now, make use of Equation (3.7) for the special case $n = 2$ and $r = 1$. Then

$$2C1 = 1C1 + 1C0, \tag{3.8}$$

that is (Table 3.1), $2C1$ is the sum of the number directly above it and the number above and to the left. From Table 3.2 and Equation (3.8), we deduce $2C1 = 1 + 1 = 2$. Inserting this into Table 3.2, we obtain

1	1					
1	2	1				
1	$3C1$	$3C2$	1			
1	$4C1$	$4C2$	$4C3$	1		
1	$5C1$	$5C2$	$5C3$	$5C4$	1	
1	$6C1$	$6C2$	$6C3$	$6C4$	$6C5$	1

Table 3.3

Now let's work on the third row. Beginning with

$$3C1 = 2C1 + 2C0,$$

we see (Table 3.1) that $3C1$ is the sum of the number directly above it and the number above and to the left. Putting this together with the values already determined in Table 3.3, we find that $3C1 = 2 + 1 = 3$. Similarly,

$$3C2 = 2C2 + 2C1$$

$$= 1 + 2$$

$$= 3.$$

Hence, we can determine all the values in row 3 and obtain

1	1					
1	2	1				
1	3	3	1			
1	4C1	4C2	4C3	1		
1	5C1	5C2	5C3	5C4	1	
1	6C1	6C2	6C3	6C4	6C5	1

Table   3.4

By the same token, we can now fill in the entries of row 4, then of row 5, etc.

1	1					
1	2	1				
1	3	3	1			
1	4	6	4	1		
1	5	10	10	5	1	
1	6	15	20	15	6	1

Table   3.5

Table 3.5 contains the first six rows of Pascal's Triangle. Before going on, write down the next four rows (rows 7–10). This is half of Exercise 1.

There are several ways to use the computer to produce Pascal's Triangle. The most natural of these involves the new BASIC concept of an "array." We have already seen how to use "subscripted" variables like A(1), X(I), and so on. These constitute examples of 1-dimensional arrays. A 2-dimensional array involves two subscripts, for example, A(1,2), X(I,J), and BY(3,K). An array is just a variable with one or more subscripts. In the following program, we use the doubly sub-scripted symbol C(N,R) to denote the binomial coefficient nCr. The DIMension statement for arrays is a natural extention of that for sub-scripted variables. (See line 30 below.) To avoid a REDIM'D ARRAY

ERROR message, arrange your program so that the DIM statement is encountered only once. ENTER and RUN this program.

```
 10  PRINT "HOW MANY ROWS OF PASCAL'S"
 20  INPUT "TRIANGLE DO YOU WANT";M
 30  DIM C(M,M)
 40  FOR N=1 TO M
 50  C(N,0)=1:C(N,N)=1
 60  NEXT N
 70  PRINT 1;1
 80  FOR N=2 TO M
 90  PRINT 1;
100  FOR R=1 TO N-1
110  C(N,R)=C(N-1,R-1)+C(N-1,R)
120  PRINT C(N,R);
130  NEXT R
140  PRINT 1
150  NEXT N
160  END
```

Exercises (3.6)

1. By hand, without using the program, write out rows 7–10 of Pascal's Triangle. Compare the 8th row with your solution to Exercise 5, Section 3.4. (*Hint:* From Equation (3.6), each row begins and ends with a 1. From Equation (3.7), the number in row n, column r, is the sum of the number directly above it and the number above and to the left.)

2. Prove Equation (3.6).

3. Give an algebraic proof of Equation (3.7).

4. Confirm that the sum of the numbers in row n of Pascal's Triangle is 2^n for $n = 1, 2, \ldots, 10$.

5. Compute 11^2, 11^3, 11^4, ... Explain what the answers have to do with Pascal's Triangle.

6. Give a formula for the k-th term of the sequence indicated below. (*Hint:* See Exercises 10 and 12, Section 2.3.)

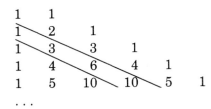

7. Form a number sequence by doing the indicated additions. Where have you seen the resulting numbers before?

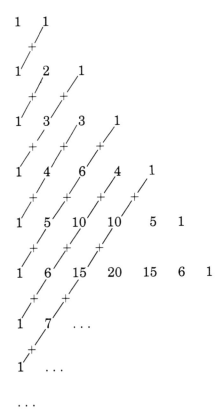

8. Draw a flowchart for the Pascal Triangle program.

9. Give a combinatorial proof that $(2n)Cn$ is always an even integer.

10. Write a program to display a "triangle" based on the numbers nPr, that is, to display the values of

$$
\begin{array}{lll}
1P1 & & \\
2P1 & 2P2 & \\
3P1 & 3P2 & 3P3 \\
\text{etc.} & \text{etc.} & \text{etc.}
\end{array}
$$

11. RUN your program from Exercise 10 to produce 7 rows of the triangle. Can you detect a pattern that would allow you to write down the 8th row from the 7th (without recourse to the algebraic formula for computing nPr)?

12. Write a program to compute Pascal's Triangle based on $nCr = nPr/r!$. (*Hint:* Modify the program you wrote for Exercise 10. This program should not use the recursion relation in Equation (3.7)).

13. One problem with outputting Pascal's Triangle to the screen is that, very soon, the rows of the Triangle become too long to fit on one line. You can improve things by omitting the first column of 1's and by eliminating the duplication resulting from the identity $nCr = nC(n-r)$. Write a program to compute and display

$$
nC1 \quad nC2 \quad \ldots \quad nCk
$$

in row n, for $n = 2, 3, \ldots$, where $k = \text{INT}((n+1)/2)$.

14. Imagine yourself mixing batter for 12 dozen chocolate-chip cookies. When you go to the cupboard for the chocolate-chips, you find someone has eaten most of them. In fact, only 300 remain. If you mix these into the batter (so that each of the 144 cookies is equally likely to get any one of the 300 chocolate chips), how many of the cookies do you expect will wind up with no chocolate chips? Write a computer program to simulate this situation. Begin by INPUTting N, the number of chocolate chips which are available to be used (in our case 300). Arrange for the computer to PRINT out a 12-by-12 array so that the entry, say, in the 3rd row and 5th column is the number of chocolate-chips which wound up in the 5th cookie of the 3rd dozen. Finally, PRINT out the total number of cookies which received 0, 1, 2, ..., 7 chocolate-chips.

3.7   EXPECTATION

As we have seen, one way to interpret probability is this: The probability of rolling, for example, a 4 with a fair die is 1/6. If a fair die is rolled 600 times, we would expect to see about 100 4's. Let's make this observation a little more formal: Suppose we repeat an experiment, like rolling a die, N times. Assume that the probability of a particular outcome is P. Then the "Expected Number" of occurrences of that outcome is $P*N$. In the case of rolling a 4 on a single die, $P = 1/6$, while $N = 600$. Hence, the Expected Number of 4's is $(1/6)*600 = 100$. It is not very likely that we would see exactly 100 4's, but, among all possible numbers of 4's (from 0 to 600), 100 is the most likely. This "most likely" number is the Expected Number.

One interesting science fiction speculation involves the existence of "extra sensory perception" or ESP. A common "test" for ESP capabilities involves a deck of 25 cards, five each of five different types (circle, square, star, plus, and waves as in Figure 3.5). A person being tested is asked to identify one of the cards as it lies face down on the table. Presumably a person endowed with this "extra" sense should consistently be able to detect the type of the card indicated. Of course, if anyone had been found who could do this (consistently), it would be big news, and you would have heard about it. Apologists for ESP suggest that the extra sense may be latent, and that those who possess it in some small degree, might be able to develop it further. How, then, can it be detected? Surely, a person with ESP should, at the very least, be able to correctly "guess" more than the Expected Number of cards. The probability of correctly guessing a card, randomly, is 1/5. Thus, in an experiment involving 100 repetitions, the Expected Number is 20. According to the enthusiasts, anyone who correctly guesses more than 20 of the cards has ESP to some degree. Not only that, but anyone

Figure   3.5

who correctly guesses fewer than 20 of the cards has a *negative* man-ifestation of ESP! It follows (if you buy this argument) that nearly everyone possesses some degree of ESP!

A concept closely related to "Expected Number" is "Expected Value." Suppose the local PTA holds a drawing to raise money. The prize is a color television worth $600. Exactly 1000 tickets will be sold for $1 each. The "Expected Value" of a ticket is the probability (P) of winning the prize times its value (V). In this case, the Expected Value of a PTA ticket is (1/1000)*600, or 60 cents. Here is how you should interpret "Expected Value." If you were to participate in a large number of such drawings, your average winnings would be 60 cents; for every dollar you spent, you would receive back, on the average, 60 cents. Of course, you would never win 60 cents. The value of your winnings would either be $600 or nothing. In the long run, however, your winnings would come close to 60 cents on the dollar.

Generally, a game (or drawing) is "fair" if the price of playing (the cost of a ticket) is equal to its Expected Value. The PTA drawing is not "fair," but no one expected it to be. In addition to the Expected Value of the ticket, there is the social "value" of contributing to the PTA fundraiser. Still, it is worth knowing that "on the average" 40 cents out of every ticket you buy should be considered an outright gift, while, in the long run (that is, if you "play the game" long enough), you may expect to get 60 cents back.

Suppose the PTA drawing was so successful that it was decided to repeat it the following year. This time, there are 3000 tickets priced at $2 each. Instead of just one prize, there are six. The grand prize is a $500 color TV. In addition, there are to be five $200 black-and-white TV's given away as "first" prizes. The Expected Value of a $2 ticket in this drawing is

$$(1/3000)*500 + (5/3000)*200 = .50.$$

This represents the probability of winning the color TV times its value, plus the probability of winning one of the black-and-white TV's times its value. How does this compare with the previous drawing? Last year you paid $1 for a chance to win 60 cents. This year, you're being asked to pay $2 for a chance to win 50 cents. (Which is the better deal for you? For the PTA?)

One popular gambling game at charity fundraisers is "Chuck-a-Luck." The apparatus for the game consists of three dice in an hourglass shaped cage. When a customer pays \$1, the operator turns the cage over and the dice roll to the bottom. If a single 5 comes up, the customer receives \$2. If two 5's come up, the prize is \$3. If three 5's come up, the player receives \$4. But, if no 5's show, the customer simply loses the dollar it cost to play the game. Let's compute the Expected Value of playing Chuck-a-Luck. By the Fundamental Counting Principle, there are 6*6*6 = 216 equally likely outcomes for the three dice. Of these, only one consists of three 5's. Thus, the probability of winning \$4 is 1/216. What is the probability of exactly two 5's showing? Think of it this way. There are two decisions to be made. The first is which die should show some digit other than 5. The second decision is what should come up on this die. There are three choices for the first decision and five for the second. Thus, there are 3*5 = 15 ways for the three dice to show exactly two 5's. What about just one 5? Decision 1 is which of the three dice should bear the 5 (three choices). The second decision is what to have come up on the second die (five choices, anything but a 5). The third decision is what to have come up on the third die (five choices). So, there are 3*5*5 = 75 ways to produce exactly one 5. So far, we have accounted for 1 + 15 + 75 = 91 out of the 216 possible outcomes. (In other words, the probability of rolling *at least* one 5 when three dice are tossed is 91/216.) The remaining 216 − 91 = 125 (equally likely) possibilities involve no 5's at all. It is in these cases that the customer walks away empty.

Number of 5's	Pay-off	Probability
0	0	125/216
1	2	75/216
2	3	15/216
3	4	1/216

The Expected Value for Chuck-a-Luck is

$$0*(125/216) + 2*(75/216) + 3*(15/216) + 4*(1/216)$$

$$= (0 + 150 + 45 + 4)/216$$

$$= 199/216,$$

or 92.1 cents. (On the average, you can expect to lose 7.9 cents every time you play Chuck-a-Luck.)

If you have extra time, here is a game you can ENTER and play:

```
 10 PRINT "PLAY CHUCK-A-LUCK"
 20 PRINT
 30 PRINT "PRESS 1 TO PLAY, 0 TO QUIT."
 40 INPUT A
 50 IF A<>1 THEN 600
 60 D1=1+INT(6*RND(1)):PRINT D1;
 65 D2=1+INT(6*RND(1)):PRINT D2;
 70 D3=1+INT(6*RND(1)):PRINT D3;
 80 IF D1=5 AND D2=5 AND D3=5 THEN 200
 90 IF D1=5 AND D2=5 AND D3<>5 THEN 300
100 IF D1=5 AND D2<>5 AND D3=5 THEN 300
110 IF D1<>5 AND D2=5 AND D3=5 THEN 300
120 IF D1<>5 AND D2<>5 AND D3<>5 THEN 400
130 PRINT " YOU WIN $2.":GO TO 20
200 PRINT " YOU WIN $4.":GO TO 20
300 PRINT " YOU WIN $3.":GO TO 20
400 PRINT " YOU WIN NOTHING.":GO TO 20
600 END
```

The purpose of line 20 is simply to instruct the computer to PRINT a blank line—to "double space." Of course, lines 20–30 could be crunched into the single line

```
20 PRINT:PRINT "PRESS 1 TO PLAY, 0 TO QUIT."
```

Of more interest, perhaps, is the use of the "logical connective" AND in lines 80–120. Take line 80, for example. In order for the computer to reach the THEN part, all three criteria must be met. To the computer, AND is an "all-or-nothing" instruction. (Compare with the use of OR in Section 2.4.). Finally, the "< >" symbol means "not equal."

Exercises (3.7)

1. Suppose you were to take a True/False test consisting of 50 questions, *without reading the questions!* "Reach out with your feelings"; use "THE FORCE." Or, use ESP. If your answers turn out to be random, after all, what is your Expected Number of correct answers? What grade do you think such a score should receive?

2. In California, the written test for a driver's license consists of 36 multiple-choice questions (3-choice questions, in fact).
   a. How many different ways are there to answer (all the questions on) such a test?
   b. What is the Expected Number of correct answers if the bubbles are filled in at random?

3. On Saturday night, August 4, 1984, "lotto fever" peaked in Ohio. A U.S. record jackpot of nearly $25 million was at stake. To win the jackpot, you would have needed to pick the correct 6 numbers from a total of 40.
   a. What is the probability of picking the correct 6 numbers?
   b. If a $1 ticket buys two plays (that is, gives you the opportunity to make two 6-number selections), what is the Expected Value of a ticket? (Assume that no two people pick the winning combination.)

4. "Mortality tables" are used by the life insurance industry to set premiums on policies. Here is a portion of a table for ages 15–25.

Age	P
15	.0015
16	.0016
17	.0016
18	.0017
19	.0017
20	.0018
21	.0018
22	.0019
23	.0019
24	.0019
25	.0019

In this table, P is the empirical probability that death will occur in the next year. If a $10 thousand policy is issued on a 16-year-old, what portion of the premium should the company place in reserve to pay death benefits? (*Hint:* This is an Expected Value problem.)

5. If the other payoffs remain the same, what should the payoff for three 5's be in order to make Chuck-a-Luck a fair game?

6. There is a version of Chuck-a-Luck in which the gambler is allowed to choose his/her favorite number (from 1 to 6) before playing the game. In this version, the "house" pays if the three dice show the favorite number at least once. Write a computer program to simulate this version of Chuck-a-Luck.

7. If you flip a (fair) coin 100 times, the probability of obtaining exactly 50 heads and 50 tails is $(100C50)/(2^{100})$.
   a. Explain why.
   b. Compute this probability to 6 decimal places. (*Hint:* Use the computer, but don't expect it to evaluate 100!.)

8. In the late 1700s, St. Petersburg (now Leningrad) was one of the foremost centers of mathematical activity. Catherine the Great was an enthusiastic patron of the arts and sciences. (In which group would you say mathematics belongs?) Among the people Catherine attracted to her court were Nicholas and Daniel Bernoulli, members of the justly famous family of Swiss mathematicians. Associated with this time and place is a famous problem, attributed to the Bernoullis, which has come to be known as the St. Petersburg Paradox.

   Imagine the following situation: You are attending a "Computer Fair." One of the attractions is a new microcomputer being introduced by a major company. The computer is programmed to simulate a succession of coin flips. The program works this way. As long as the simulated flip is "heads," the computer "flips" again. As soon as a "tail" turns up, the program ENDs. Suppose you are afforded the opportunity to participate in a "game" with the computer. If the program ENDs after producing one head, you get $1. If the program ENDs after producing two heads you get $2, after three heads, you get $4, after four heads, you get $8, . . . , after n heads, you get 2^{n-1} dollars. What should you be willing to pay for the privilege of playing this game? (What is the Expected Value of the game? You get nothing if the program ENDs before it produces any heads.)

9. Write the program for the new microcomputer demonstration (attraction) discussed in Exercise 8.

10. Modify the program that you wrote in Exercise 9 so that it will display the results of 100 simulations of the demonstration and keep track of the number of times the program ENDed after flipping 0 heads, 1 head, ..., 9 heads, and 10-or-more heads. RUN this program five times and record the result.

11. What is the "paradox" in the St. Petersburg Paradox?

12. A Nevada-style $1 slot-machine has three wheels that rotate independently. Suppose the pictures on each wheel occur with the frequencies shown. Assume the prizes are $10 for three Cherries, $8 for three Oranges, $9 for three Plums, $20 for three Bells, and $50 for three Bars.

Picture	Wheel		
	1	2	3
Cherry	2	5	6
Orange	5	3	5
Plum	5	5	3
Bell	5	3	2
Bar	1	2	2
Lemon	2	2	2

    a. How many equally likely outcomes are there for this machine?
    b. What is the probability of "hitting" the $50 Jackpot?
    c. What is the Expected Value of the machine?
    d. If the other prizes remain the same, what should the Jackpot pay for the machine to be "fair"?

13. Write a program to simulate a play of the slot-machine in Exercise 12. (You needn't program the machine to announce the prize.)

14. Explain the difference between each of the following:
    a. `PRINT 1 + INT ( 6 * RND ( 1 ) ) + 1 + INT ( 6 * RND ( 1 ) )`
    b. `PRINT 2 * ( 1 + INT ( 6 * RND ( 1 ) ) )`
    c. `PRINT 1 + INT ( 12 * RND ( 1 ) )`

3.8   THE PROBABILITY OF *A* OR *B*

Before continuing our development of probability theory, let's pause to discuss some programming "twists." Up to now, we haven't been too concerned about how/when/where output appears on the screen. Arranging the output in a particular way is referred to as "formatting." It is possible, for example, to incorporate CuRSoR movement into a program. Suppose you want to leave five spaces between two lines of output on the screen. One way to do it is

```
   . . .
110 PRINT [FIRST LINE]
120 PRINT
130 PRINT
140 PRINT
150 PRINT
160 PRINT
170 PRINT [SECOND LINE]
   . . .
```

If the dialect of BASIC "spoken" by your computer supports crunching, you could simplify the procedure to

```
   . . .
110 PRINT [FIRST LINE]
120 PRINT:PRINT:PRINT:PRINT:PRINT
130 PRINT [SECOND LINE]
   . . .
```

or do something like

```
   . . .
110 PRINT [FIRST LINE]
120 FOR I=1 TO 5:PRINT:NEXT I
130 PRINT [SECOND LINE]
   . . .
```

The new idea is to replace line 120 in the second and third of these alternatives with

```
120 PRINT "[CRSR down - 5 times]"
```

In this line, [CRSR down—5 times] means activate the CRSR down procedure five times. This may simply be a matter of pressing a CRSR key; it may involve PRINTing a particular code word; it may involve a special procedure peculiar to your machine and its keyboard; or, it may not be possible at all. In some versions of BASIC, putting a CRSR movement instruction inside quotes in a PRINT statement is a way to put CRSR movement into a program. Experiment with this procedure and determine the extent to which it works on your machine. (Other possibilities include PRINT "[CLeaR (screen) key]", PRINT "[HOME key]", PRINT "[CRSR up/left/right key(s)]", etc.† A combination of CLR and HOME‡ is a good way to begin many programs.)

Another "twist" involves introducing some delay into your programs. If, for example, you want to simulate a sequence of coin flips, and make it look real, you can introduce a line like

```
200 FOR T=1 TO 100:NEXT T
```

to create a (slight) time delay. If you want a longer delay, use T = 1 to 1000. ENTER line 200 above and see how long it takes to RUN.

Returning to our theoretical discussion, suppose the probability of some event is P. Then, of course, $0 \leqslant P \leqslant 1$. If $P = 0$, the event won't occur and if $P = 1$, it is certain to occur. (We will see later that P = 0 does not necessarily mean that the event can't occur! If the number of possible, equally likely events is infinite, "can't occur" may be different from "won't occur.")

Given that a particular event occurs with probability P, the probability that it won't occur is $1 - P$. In the case of finitely many, say d, equally likely events, n of which are noteworthy, $P = n/d$. The number of non-noteworthy events is $d - n$ so the probability that one of *them* will occur is $(d - n)/d = 1 - (n/d) = 1 - P$. (The argument in the infinite case is a little more sophisticated; it won't be given here.)

Before going on, we need to introduce some new notation. If A is some event, denote by $P(A)$ the probability that it will occur. Then, by what we have just said, $1 - P(A)$ is the probability that it won't occur. For example, if a fair die is rolled, let T stand for the event that the die shows a number less than 3. Then $P(T) = 2/6 = 1/3$. The probability that T doesn't occur, that is that the die shows a number at least 3, is $4/6 = 1 - P(T)$.

---

† It may be that your version of BASIC supports none of these "frills."

‡ HOME is the upper left hand corner of the screen.

Let us examine this problem from another perspective. If the die shows a number less than 3, then it shows 1 or 2. Thus,

$$P(1 \text{ or } 2) = 2/6$$
$$= 1/6 + 1/6$$
$$= P(1) + P(2).$$

If A is the event that a 1 occurs and B is the event that a 2 occurs, then $P(A \text{ or } B) = P(A) + P(B)$. Does this formula hold in general? The answer is no. The reason it is "no" involves us in another counting principle. Consider the two sets, X and Y, in Figure 3.6. Letting $o(X)$ stand for the number of elements in X, let's determine $o(X \cup Y)$, the number of elements in their union. If we take $o(X) + o(Y)$, we have too much. We will have counted the number of elements in $X \cap Y$ twice. But, we can make up for this "double dipping" by subtracting $o(X \cap Y)$. Thus, we have a *Second Counting Principle*:

$$o(X \cup Y) = o(X) + o(Y) - o(X \cap Y) \qquad (3.9)$$

Let the set E in Figure 3.6 represent the totality of all equally likely outcomes.[†] Then the probability of an event of type X occurring is $P(X) = o(X)/o(E)$. Similarly, the probability of an event of type Y occurring is $P(Y) = o(Y)/o(E)$. So, using (3.9),

$$P(X \text{ or } Y) = o(X \cup Y)/o(E)$$
$$= P(X) + P(Y) - o(X \cap Y)/o(E)$$

How do we interpret the last term? It looks like a probability, having $o(E)$ in the denominator. The numerator is the number of elements of E of type X *and* type Y. Thus,

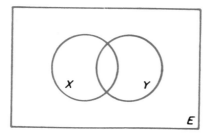

Figure 3.6

[†] The set E is sometimes called a "Sample Space."

$$P(X \text{ or } Y) = P(X) + P(Y) - P(X \text{ and } Y) \qquad (3.10)$$

We say two outcomes are *disjoint* if they cannot occur simultaneously. If A and B are disjoint outcomes, then $P(A \text{ and } B) = 0$ and

$$P(A \text{ or } B) = P(A) + P(B) \qquad (3.11)$$

Suppose, for example, that a single die is rolled. Let A be the outcome that the die shows a 4, and B the outcome that the die shows a 5. These two outcomes are disjoint since they cannot occur at the same time. If, on the other hand, A is the outcome that the die shows an odd number and B is the outcome that the die shows a number no more than 3, then the outcomes are not disjoint. In this case $P(A) = 1/2 = P(B)$. To compute $P(A \text{ or } B)$, note that the favorable outcomes are 1, 2, 3, or 5. Thus, $P(A \text{ or } B) = 4/6 = 2/3$. What about $P(A \text{ and } B)$? Now we're talking about an odd number no more than 3. There are two such numbers, namely, 1 and 3. Thus, $P(A \text{ and } B) = 2/6 = 1/3$.

Exercises (3.8)

1. Explain how to accomplish each of the following screen editing tasks within a program. (If you were not able to accomplish one or more of the tasks, say so.)
   a. Move the CRSR up.
   b. Move the CRSR down.
   c. Move the CRSR to the right.
   d. Move the CRSR to the left.
   e. HOME the CRSR.
   f. CLeaR the screen.

2. How long does it take to RUN this program?

```
200 FOR T=1 TO 10000:NEXT T
210 END
```

3. Based on Exercise 2, how long would you expect this line to delay the computer?

```
200 FOR T=1 TO 1000:NEXT T
```

4. Check your estimate in Exercise 3 by timing a RUN or two of this program.

```
200 FOR T=1 TO 1000:NEXT T
210 END
```

5. Write a program to have the computer PRINT your name on the screen, a letter at a time, from right to left.

6. If we denote a "favorable" outcome by A, let A' represent an "unfavorable" outcome. Show that $P(A) = 1 - P(A')$ is a special case of one of the formulas for $P(A$ or $B)$.

7. If one die is rolled, which of the following are disjoint outcomes?
   a. $A = 3$ shows; $B = 4$ shows.
   b. $A = 3$ shows; $B =$ odd shows.
   c. $A = 3$ shows; $B =$ even shows.
   d. $A = 3$ shows; $B =$ anything shows.
   e. $A =$ even shows; $B =$ odd shows.
   f. $A = 2$ or 4 show; $B = 2$ or 6 show.

8. We say that three outcomes, A, B, and C, are disjoint if no two of them can hold simultaneously. Which of the following are disjoint outcomes?
   a. $A = 3$ shows; $B =$ even shows; $C = 1$ or 5 show.
   b. $A =$ a number $\leqslant 3$ shows; $B =$ even shows; $C =$ odd shows.
   c. $A =$ nothing shows; $B =$ odd shows; $C = 2$ shows.
   d. $A = 1$ shows; $B = 2$ shows; $C = 3$ shows.

9. If A, B, and C are disjoint outcomes (See Exercise 8.), show that $P(A$ or B or $C) = P(A) + P(B) + P(C)$.

10. Compute the probability, $P(A$ or $B)$, for each part of Exercise 7.

3.9 CONDITIONAL PROBABILITY

In some IF ... THEN situations, it is convenient to phrase the IF condition in terms of one variable, X, being unequal to a second variable, Y. This is done as follows:

```
110 IF X<>Y THEN ...
```

The "<>" symbol is a combination of "<", "less than," and ">", "greater than." It is interpreted by the computer to mean "not equal," that is, "$\neq$". In a similar way, "$<=$" represents "$\leq$", that is, "less than or equal to," while "$>=$" is interpreted as "$\geq$".

There is another code word that is occasionally used with IF . . . THEN. It is ELSE, as in

```
310  IF  X=Y  THEN  200  ELSE  50
```

The idea is that if $X<>Y$, then execution should transfer to line 50. Generally speaking, this construction is avoided as it makes the program difficult (for humans) to follow.

In the last section, we discovered that

$$P(A \text{ or } B) = P(A) + P(B) - P(A \text{ and } B).$$

Let's give some attention to the problem of computing $P(A \text{ and } B)$. Consider this example: Suppose you roll two dice, a red one and a green one. Let A represent the outcome that the number on the red die is < 3. Denote by B the outcome that the green die comes up even. Then what is $P(A \text{ and } B)$, the probability that both A and B occur?

The total number of equally likely possibilities is $6*6 = 36$. To compute the number of these that are noteworthy, we appeal to the Fundamental Counting Principle: The first (red) decision involves two choices, while the second involves three. Thus, the number of noteworthy events is $2*3$ and the desired probability is

$$P(A \text{ and } B) = (2*3)/(6*6)$$

$$= (2/6)*(3/6),$$

the product of $P(A)$ and $P(B)$.

In many cases, the probability that A and B will both occur is $P(A)*P(B)$. Consider another example. Suppose you and two friends just happen to be in a local convenience store when a soft drink distributor is conducting a promotional campaign. The distributor presents you and your friends with the following opportunity: Without collaborating, you should each choose a number between 1 and 6. If you all choose different numbers, you each win a six-pack of your favorite soft drink (12-oz. cans). If, however, two of you should both pick the same number, you win nothing. What is the probability that the three of you will win?

To make the computation easier, suppose you choose and write down your number first (and maybe try to use ESP to signal your friends which number to avoid). Perhaps friend G writes down her number second, and friend B writes his down last. Let's compute the number of favorable outcomes. There are three decisions to be made. The first (yours) involves six choices. The second involves five, and the third only four. Thus, the probability that the three of you win is

$$(6*5*4)/(6*6*6) = (5/6)*(4/6).$$

The probability that friend G avoids your number is 5/6 and the probability that friend B avoids both numbers is 4/6 or 2/3. Thus, the probability that you and your friends win is $(5/6)*(2/3) = 5/9$. (Does it follow from this argument that friend B has the biggest responsibility? No, but why not?)

Here is another way to arrive at the answer. Look back at the Chuck-a-Luck Table in Section 3.7. (The three of you might just as well have made your choices by rolling three dice!) There are 16 ways for three dice to show at least two 5's. So, there must be 16 ways for the dice to show at least two 1's, 16 ways to show at least two 2's, etc. In other words, there are $6*16 = 96$ ways to lose the distributor's contest. It follows that there must be $216 - 96 = 120$ ways to win. Hence, the probability of winning is $120/216 = 5/9$.

It is not always true that $P(A \text{ and } B)$ is the product of $P(A)$ and $P(B)$. In general, we need to be just a little more sophisticated. Denote by $P(B \mid A)$ the probability of B given that A has already occurred. Then $P(B \mid A)$ is sometimes called a *conditional probability*. It is the probability of B given the condition that A has already happened.

THEOREM:

If A and B are two outcomes, then

$$P(A \text{ and } B) = P(A)*P(B \mid A) \qquad (3.12)$$

Proof. Suppose the total number of equally likely possibilities is d, and that the number of these of type A is a. Then $P(A) = a/d$. If, among the a events of type A, there are b of type B, then $P(A \text{ and } B) = b/d$, while $P(B \mid A) = b/a$. But then the product on the right hand side of Equation (3.12) is $(a/d)*(b/a) = b/d$. Q.E.D.

Say that events A and B are *independent* if $P(B \mid A) = P(B)$. That is, A and B are independent if the probability of B is the same whether A has occurred or not. Notice that Equation (3.12) takes the form $P(A \text{ and } B) = P(A)*P(B)$ precisely when A and B are independent events.

There is a slightly troublesome point about Equation (3.12). While the left-hand side is clearly symmetrical, the right-hand side seems less so. However, if we just interchange A and B, we obtain

$$P(B \text{ and } A) = P(B)*P(A \mid B) \qquad (3.13)$$

Comparing Equations (3.12) and (3.13), and taking into account that $P(A \text{ and } B) = P(B \text{ and } A)$, we obtain Bayes's First Rule:[†]

$$P(A)*P(B \mid A) = P(B)*P(A \mid B) \qquad (3.14)$$

In particular, $P(B \mid A) = P(B)$ if and only if $P(A \mid B) = P(A)$.

In conclusion, let's discuss the difference between "independent" events and "disjoint" outcomes. (In a sense, the two concepts are opposites.) If we draw one card from a 52-card deck, it might be: A—a king; or B—a club. These outcomes are independent: $P(B) = 13/52 = 1/4 = P(B \mid A)$. The probability that the card is a club is independent of whether we have prior knowledge that it is a king. (Note that $P(A)*P(B) = (1/13)*(1/4) = 1/52 = P(\text{king of clubs}) = P(A \text{ and } B)$.) On the other hand, the two outcomes, A and B, are not disjoint since they can occur simultaneously. Indeed, $P(A \text{ or } B) = 16/52$ while $P(A) + P(B) = 4/52 + 13/52 = 17/52$.

Another possibility is that the card is C—a queen. The outcomes, A and C are disjoint: the card cannot be both a king and a queen. But, A and C are not independent. The probability that the card is a queen is clearly affected by whether we know it to be a king! $P(C) = 1/13$ while $P(C \mid A) = 0$.

Finally, consider the possible outcome that the card is D—a face card (that is, a king, queen, or jack). Outcomes A and D are neither independent nor disjoint. Certainly, they can occur simultaneously. Moreover, $P(D) = 3/13$, while $P(D \mid A) = 1$.

Exercises (3.9)

1. If the following program were RUN, what would the output be? (This is an exercise for you, not for the computer.)

[†]Thomas Bayes (1702–1761) was an English mathematician and clergyman. He defended Newton's Calculus against the philosophical attack of Bishop Berkeley. He is best known, however, for his *Essay Towards Solving a Problem in the Doctrine of Chances*.

```
 10  X=7:Y=13
 20  IF  X>=Y  THEN  40  ELSE  50
 30  GO  TO  60
 40  PRINT  "PLEASE"
 50  PRINT  "GO";
 60  IF  X<>Y  THEN  80
 70  GO  TO  40
 80  IF  X<=Y  THEN  110
 90  PRINT  "HOME."
100  GO  TO  120
110  PRINT  "OD"
120  END
```

2. If two dice (red and green) are rolled, which of the following are independent events?

   a. $A = 3$ on the red die; $B = 3$ on the green die.
   b. $A = 3$ on the red die; $B =$ odd number on the green die.
   c. $A = 3$ on the red die; $B =$ total at least 4 on both dice.
   d. $A = 3$ on the red die; $B =$ odd total on both dice.
   e. $A = 3$ on the red die; $B =$ total at least 3 on both dice.
   f. $A = 3$ on the red die; $B =$ total at least 2 on both dice.

3. Compute $P(A \mid B)$ and $P(B \mid A)$ for each part of Exercise 2.

4. Confirm Equation (3.14) for each part of Exercise 2.

5. Compute $P(A \text{ and } B)$ for each part of Exercise 2. (*Hint:* Use Equation (3.12) or (3.13).)

6. In 1654, Antoine Gombaud, the Chevalier de Méré was interested in a dice game in which he bet that a 6 would appear at least once when four dice were rolled. What is the probability that de Méré won in any particular instance of this game? (*Hint:* First compute the complementary probability that no 6's appear.)

7. Perhaps because he could no longer find anyone to take his bet (see Exercise 6), the Chevalier switched to betting that "boxcars" (double 6's) would occur at least once in any 24 consecutive rolls of two dice. What is the probability that he won in any particular instance of this new game?

8. The Chevalier de Méré may have believed that the probability of winning is the same in each of the games he played. (See Exercises

6–7.) Discuss the following line of reasoning: "In one roll of a single die, the chance of a 6 is 1/6. So, if four dice are rolled, my probability of winning is 4*(1/6) = 2/3. On the other hand, in one roll of a pair of dice, the chances of rolling a double 6 are 1/36. So, if the dice are rolled 24 times, my probability of winning is 24*(1/36) = 2/3."

9. Suppose you toss a half-dollar coin n times. How large must n be to guarantee that your odds of tossing a head at least once are 100-to-1. (Remember, odds of 100-to-1 are equivalent to a probability of 100/101.)

Let's generalize the promotional give-away problem. Suppose k people independently choose 1 number from a total of m possible numbers. (In the case of you and your two friends, $k = 3$ while $m = 6$.) What is the probability that no two people choose the same number? So that the solution will be nontrivial, assume $k < m$. Imagine the k choices made in succession. After the first person has chosen a number, the probability that the second person will choose a different number is $(m-1)/m$. If person 2 succeeds in avoiding a duplication, the probability that the third person picks an unchosen number is $(m-2)/m$, and so on. Finally, if the first $k-1$ people have managed to make different selections, the probability that person k will pick an (as yet) unchosen number is $(m-k+1)/m$. So, the probability that no two people will pick the same number is the product,

$$P = [(m-1)/m]*[(m-2)/m]* \cdots *[(m-k+1)/m]$$ (3.15)

$$= mPk/(m^k)$$

Turning the problem around, if k people select a single number from 1 up to m, what is the probability that some (at least one) number will be chosen more than once? It is $1-P$, where P is the probability calculated in Equation (3.15).

10. Suppose k people independently choose one number from 1, 2, . . . , m. Let P be the probability that they all choose different numbers. Write a program to INPUT m and k, and to output P.

11. RUN the program you wrote in Exercise 10, and list the results for each of the following pairs m, k. (Round off to two decimal places.)
    a. 6, 3   b. 10, 3   c. 100, 15   d. 365, 40   e. 1000, 50

12. Suppose k people independently choose one number from $1, 2, \ldots,$ m. Let Q be the probability that some two of the people choose the same number. Compute Q for each of the following pairs m, k. (*Hint:* Exercise 10. Round off to 2 decimal places.)

    a. 6, 3   b. 10, 3   c. 100, 15   d. 365, 40   e.1000, 50
    f. 100, 21   g. 21, 100   h. 20, 10   i. 100, 50

13. Suppose 40 people are chosen at random from a crowd. What is the probability that some pair of them share the same birthday? (Assume that none of them was born on February 29. See Problem 12d.)

14. Suppose you are one of the 40 people in Exercise 13. What is the probability that (at least) one of the 39 other people shares your birthday?

15. How many ways are there to choose a pair of people from 40 people? Explain what, if anything, this has to do with the previous two exercises.

3.10 BINOMIAL PROBABILITY

If you did Exercises 13–15 in Section 3.9, you encountered the so-called "Birthday Paradox." In any random group of 40 people, the probability is nearly .9 that (at least) two of them share the same birthday. This is a paradox only in the sense that it seems to defy our intuition. One way to convince yourself that the calculation is correct is to interview many groups of 40 people, say at a sporting event. An easier approach would be to use the computer to simulate such an experiment. Here is a program to do just that. RUN it and use the output to compute an empirical probability. (Choose N to be at least 20.)

```
  10 REM PROGRAM TO SIMULATE THE BIRTHDAY
  15 REM PARADOX—40 PEOPLE—365 DAYS
  20 DIM B(40):C=0
  30 INPUT "HOW MANY TRIALS";N
  40 FOR T=1 TO N
  45 PRINT " * * * * * * * * * * * * * * * * * * * * * * * * "
  50 FOR I=1 TO 40
  60 B(I)=1+INT(365*RND(1))
  70 PRINT B(I);
  80 IF I=1 THEN 150
  90 FOR J=1 TO I—1
 100 IF B(J)<>B(I) THEN 140
 105 PRINT
 110 PRINT "PERSONS";J;"AND";I;"HAVE";
 115 PRINT "THE SAME BIRTHDAY"
 120 C=C+1
 130 J=I—1:I=40
 140 NEXT J
 150 NEXT I
 160 NEXT T
 165 PRINT
 170 PRINT "OUT OF";N;"TRIALS, THERE WERE"
 180 PRINT C;"WITH REPEATED BIRTHDAYS."
 200 END
```

After line 120, we are ready to proceed to the next trial (having encountered two "people" with the same birthday). In other words, we are ready to exit from the I and J loops. This is accomplished in line 130 by setting the parameters I and J equal to their terminal values. We might accomplish the same thing by changing line 130 to

```
130 GO TO 160
```

One has to be a little careful about jumping out of loops before they have completed their allotted repetitions. If a loop is nested within another loop, the inner loop will be cancelled whenever the outer loop reaches a NEXT statement. If a loop is opened within a (GOSUB) subroutine, RETURN from that same subroutine will automatically scratch the loop. Or, if a FOR statement is encountered using the same variable name, any previously existing loop using the same variable name will be eliminated. Jumping from a loop under other circumstances may cause problems because the computer doesn't forget, even if you may, that it is in the middle of a loop. One kind of problem is this. A microcomputer is able to keep track of only a limited number, maybe 8–12, loops at one time. If it is called upon to remember more loops than that, it will terminate execution with an OUT OF MEMORY ERROR. This can be confusing, especially if you are not using a lot of memory. The best strategy when you exit from a loop early (and don't intend to come back) is to make a point of cancelling the loop.

In a popular game called *Yahtzee*, five dice are rolled in hopes of obtaining various combinations. Suppose you need to roll three 4's to win the game. What is the probability of rolling exactly three 4's in a single throw of the five dice?

If exactly three 4's come up, they could come up in 5C3 = 10 disjoint ways. The "first" three dice might show 4's, while the remaining two show something else; dice 1, 2 and 4 might show 4's while dice 3 and 5 show something else, and so on. Label these ten outcomes $A0$, $A1$, ..., $A9$. Then, using Equation (3.11) from Section 3.8 repeatedly, we obtain

$$P(A0 \text{ or } A1 \text{ or } \ldots \text{ or } A9) = P(A0) + P(A1) + \cdots + P(A9)$$

Since $P(A0) = P(A1) = \cdots = P(A9)$, the answer is $10*P(A0)$. Now, $P(A0)$ is the probability that the "first" three dice show 4's, while the other two dice do not show 4's. This is just

$$P(A0) = (1/6)*(1/6)*(1/6)*(5/6)*(5/6)$$

So, the answer we seek is

$$[5C3]*(1/6)^3*(5/6)^2 = .032 \ldots$$

The pattern seems clear already. The probability of rolling exactly two 4's is $[5C2]*(1/6)^2*(5/6)^3$. The probability of rolling exactly r 4's is $[5Cr]*(1/6)^r*(5/6)^{5-r}$.

More generally, *binomial probability* involves the following situation: Suppose we describe a noteworthy outcome as a success and anything else as a failure. If the probability of a success is p, than the probability of failure is $q = 1-p$. Our discussion may then be summarized as follows: The probability that we will succeed r times in n tries is

$$P(r \text{ successes in } n \text{ tries}) = [nCr]*p^r*q^{n-r}. \qquad (3.16)$$

We have already discussed the special case $p = q = .5$ in Section 3.4. An example of binomial probabilities in which $p \neq q$ occurs in *telemetry*, the science of transmitting data from remote sources. Consider, as an example, pictures taken of Jupiter by a passing unmanned vehicle. Obviously, Polaroid prints are not put in an envelope and mailed to Earth. Typically, the picture is sent as a sequence of 1's and 0's, as a "binary numeral,"† or "0-1 code." The problem is that the signal is so faint, that it can be disrupted by "noise," such as interference caused by radiation from the Sun. Suppose that a 0-1 code consisting of n digits is sent. Suppose the probability is p that any one of these digits will be incorrectly decoded, as a 1 when 0 is sent, or as a 0 when 1 is sent. What is the probability that exactly r digits are decoded incorrectly?

As before, let $q = 1 - p$. Then, the probability that the first r digits are decoded incorrectly, and rest correctly, is $p^r q^{n-r}$. Indeed, this is the probability that any particular set of r digits is decoded incorrectly. Since there are nCr, r-digit subsets of the n digits, the probability that exactly r digits are decoded incorrectly is $(nCr)\, p^r q^{n-r}$.

What is the probability that, at most, r digits are incorrectly decoded? It is the sum of $(nCk)\, p^k q^{n-k}$ as k runs from 0 up to r, that is,

$$q^n + npq^{n-1} + \cdots + (nCr)\, p^r q^{n-r}.$$

Exercises   (3.10)

1. RUN the Birthday Paradox Program in the text. INPUT N = 20 and record the result.

2. RUN the Birthday Paradox Program. INPUT N = 100 and record the result.

3. Explain the purpose of line 45 in the Birthday Paradox Program.

4. Suppose six dice are rolled. What is the probability that
   a. exactly three 4's will show?
   b. exactly five 4's will show?
   c. at least five 4's will show?
   d. no 4's will show?

5. Suppose a single die is rolled six times. What is the probability that
   a. exactly three 4's will show?
   b. exactly five 4's will show?

---

† See Chapter 5.

    c. at least five 4's will show?

    d. no 4's will show?

6. The following problem was once posed by the diarist Samuel Pepys to Isaac Newton: Who has the greatest chance of success:

    a. a man who throws six dice in hopes of obtaining at least one 6,

    b. a man who throws twelve dice in hopes of obtaining at least two 6's,

    c. or a man who throws eighteen dice in hopes of obtaining at least three 6's?

    (Give the probability of success in each case as a justification for your answer.)

7. Suppose a string of 100 0's and 1's is received by a telemetry station. Suppose the probability of correctly decoding any single digit is .995. What is the probability that exactly r errors occur in the entire string of 100 digits if $r =$

    a. 0?   b. 1?   c. 2?   d. 3?   e. 4?   f. 5?   g. 6?

8. In the telemetry situation of Exercise 7, what is the probability that six or fewer decoding errors occur in the entire string of 100 digits?

Chapter 4
STATISTICS

4.1   DATA

Here is a new way to assign values to variables in a program. It involves the BASIC code words READ and DATA. ENTER and RUN this program.

```
10  FOR I=1 TO 5
20  READ A(I)
30  PRINT A(I)
40  NEXT I
50  DATA 17,5,39,22,1
60  END
```

Let's go through this program line-by-line, as if we were the computer. At line 10, we place the number 1 in a memory location labeled I. At line 20, we label a memory location A(1). The READ statement instructs us to look for something to put into this memory location. We look for a DATA statement, which can be almost anywhere in the program but is usually found either at the beginning or the end. Since this is our first READ, we take the first number in the (first) DATA statement (namely, 17) and put *it* in memory location A(1). Moreover, we keep track of the fact that we have "read" one number. Moving on to line 30, we PRINT 17 and continue to line 40. Here, we inspect memory location I to see if it contains 5 (or more!). Since it doesn't, we loop back to line 10 and "increment" I by adding 1 to the number already in that location. Coming again to line 20, we are instructed to label a memory location A(2) and go looking for something to put in it. Since this is our second time to READ, we find the second DATA number (namely, 5) and put it into memory location A(2). At line 30, we return to memory location A(2), and PRINT 5, the number we find there. At line 40, we inspect the number in location I. Since it is less than 5, we loop back to line 10 and increment I. This process continues until I has the value 5 at line 40. Line 50 is now ignored, and the program terminates at line 60.

What happens if you make the DATA statement line 5 instead of line 50? (Nothing, but verify it.) What happens if you add a 6th number to the DATA statement and leave the rest of the program alone? (Nothing. No harm is done, but verify it.) What happens if you list only four numbers in the DATA statement? (You'll have a crabby computer on your hands. It will let you know what it thinks of the situation by giving you an OUT OF DATA ERROR message. Verify it.) Try RUNning this program.

```
10  FOR I = 1 TO 5
20  READ A ( I )
30  PRINT A ( I )
40  NEXT I
50  DATA 17
60  DATA 5 , 39
70  DATA 22
80  DATA 1
90  END
```

As you see, it doesn't matter whether you have one or several DATA lines. Generally, it is convenient to put as many numbers in a single DATA statement as will conveniently fit on a line.

It may happen that you want to use the same DATA numbers more than once. The BASIC command RESTORE tells the computer to forget that it has read any numbers. On the next READ command, it will take the first DATA number (again). Confirm this by RUNning the following:

```
10  FOR I = 1 TO 5
20  READ A
30  PRINT A
40  RESTORE
50  NEXT I
60  DATA 1 , 2 , 3
70  END
```

(Without line 40, the computer would PRINT 1, 2, and 3, and then give you the OUT OF DATA ERROR message. Confirm this.)

The BASIC language contains a mechanism that associates with each letter of the alphabet a numerical value.[†] See what happens when you ENTER this instruction

```
PRINT CHR$(65)
```

The numbers 65–90 correspond, respectively, to the letters A–Z. Confirm this by RUNning the following program.

```
10 FOR I=65 TO 90
20 PRINT CHR$(I);
30 NEXT I
40 END
```

Some other symbols have numerical values, too. For example, 32 corresponds to a space. Add this line to the previous program, and RUN the result.

```
25 PRINT CHR$(32);
```

Now, for a little fun. ENTER and RUN this.

```
10 FOR I=1 TO 28
20 READ A
30 PRINT CHR$(A);
40 NEXT I
50 DATA 78,79,84,72,73,78,71,32,87,79,82
60 DATA 84,72,32,68,79,73,78,71,32
70 DATA 73,83,32,69,65,83,89,46
80 END
```

To determine numerical correspondences for other symbols, we can use the ASC command. It is the reverse of CHR$. Try these

```
PRINT ASC("=")
PRINT ASC("$")
PRINT ASC("!")
```

Note the necessity of the quotation marks in the ASC command. Experiment further with CHR$ and ASC. If you have time, encode your

---

[†] This correspondence is known as the American Standard Code for Information Interchange, abbreviated ASCII (Pronounced "askey").

favorite (short) phrase in a program like the last one above. (You may find CHR$(13) helpful. It is the ASCII code for ENTER.)

Exercises (4.1)

1. What is the computer's response to each of the following?
   a. PRINT CHR$(77)   b. PRINT CHR$(88)
   c. PRINT CHR$(66)   d. PRINT CHR$(85)
   e. PRINT CHR$(87)   f. PRINT CHR$(75)

2. What is the computer's response to each of the following?
   a. PRINT ASC("D")   b. PRINT ASC("Q")
   c. PRINT ASC("C")   d. PRINT ASC("G")
   e. PRINT ASC("H")   f. PRINT ASC("R")

3. What is the computer's response to each of the following?
   a. PRINT CHR$(40)   b. PRINT CHR$(41)
   c. PRINT CHR$(37)   d. PRINT CHR$(14)
   e. PRINT CHR$(142)   f. PRINT CHR$(19)

4. What is the computer's response to each of the following?
   a. PRINT ASC("(")   b. PRINT ASC("#")
   c. PRINT ASC("*")   d. PRINT ASC("—")
   e. PRINT ASC(";")   f. PRINT ASC("?")

5. Describe the output produced when the following is RUN.

```
10  FOR I=1 TO 14
20  READ A
30  PRINT CHR$(A);
40  NEXT I
50  DATA 69,32,84,32,80,72,79,78,69,32
60  DATA 72,79,77,69
70  END
```

6. Write a program to produce a text of 10 lines so that each of the lines consists of eight 4-letter "words" made up of random letters.

7. Write a program like the one in Exercise 5 to produce your own mystery message.

8. It may surprise you to learn that the quotient of two random numbers is not random. Let x and y be random numbers between 0 and 1. Define $q = x/y$, and let q_1 be the first nonzero digit in the decimal expansion of q. If q were random, then the probability, $P(q_1 = k)$

would be 1/9 for $k = 1, 2, \ldots, 9$. Write a program to
a. INPUT N
b. Generate N pairs of RaNDom numbers, x and y.
c. For each pair, determine q_1, the first nonzero digit in the decimal expansion of x/y.
d. Use the results to (experimentally) determine $P(q_1 = 1)$, the probability that $q_1 = 1, P(q_1 = 2), \ldots, P(q_1 = 9)$.

9. RUN the program you wrote in Exercise 8 for N $= 10,000$, and report the result.

10. Write a program to
a. INPUT N
b. Generate N Pairs of RaNDom numbers, x and y.
c. For each pair, determine q, the quotient of the *smaller* of x and y, divided by the *larger*.
d. For each q, determine q_1, the first nonzero digit in its decimal expansion.
e. Use the results to (experimentally) determine $P(q_1 = 1)$, the probability that $q_1 = 1, P(q_1 = 2), \ldots, P(q_1 = 9)$.

11. RUN the program you wrote in Exercise 10 for N $= 10,000$, and report the result.

If you have extra time, here is a game you can ENTER and play:

You are the Captain of the Space Cruiser USS *Sutherland*. Sensors indicate the presence of an enemy frigate under cover of a cloaking device. Using Galactic Standard Coordinates for your quadrant, the enemy vessel is hiding in the sector (0,0,0) to (15,15,15). You can order an energy barrage to be targeted at any location in that sector.

```
 10  X=INT(16*RND(1))
 20  Y=INT(16*RND(1))
 30  Z=INT(16*RND(1))
 40  FOR I=1 TO 5
 50  PRINT "COORDINATES FOR BARRAGE";I
 60  INPUT A,B,C
 70  IF A=X AND B=Y AND C=Z THEN 200
 80  PRINT "SENSORS SHOW BARRAGE WAS ";
 90  IF A>X THEN PRINT "+";
100  IF A<X THEN PRINT "-";
110  IF A=X THEN PRINT "0";
120  IF B>Y THEN PRINT "+";
130  IF B<Y THEN PRINT "-";
140  IF B=Y THEN PRINT "0";
150  IF C>Z THEN PRINT "+"
160  IF C<Z THEN PRINT "-"
170  IF C=Z THEN PRINT "0"
180  NEXT I
190  PRINT "THE SUTHERLAND HAS BEEN"
195  PRINT "DESTROYED!": GO TO 220
200  PRINT "CONGRATULATIONS! YOU HAVE"
210  PRINT "DESTROYED THE ENEMY FRIGATE."
220  END
```

4.2   A CHOICE OF AVERAGES

Statistics is the mathematical discipline concerned with making sense
out of masses of (usually numerical) data. For example, what do you
make of this?

125	16	73	44	62
19	25	67	88	46
16	33	19	91	99
612	13	5	11	419

As a start, there are 20 numbers ranging from 5 to 612. (Already, we
are engaged in statistics!) If the physical arrangement of the numbers

is unimportant, we can profit from sorting them into increasing order. While it's hardly necessary to use a computer to sort 20 numbers, here is a program (based on Exercise 11, Section 2.5) to do the job.

```
  5  REM HOW MANY NUMBERS?
 10  READ K
 20  DIM N(K)
 25  REM READ K NUMBERS
 30  FOR I=1 TO K
 40  READ N(I)
 50  NEXT I
 60  IF K=1 THEN 220
 65  REM SORT THE NUMBERS
 70  FOR I=2 TO K
 75  REM WHERE TO PUT THE I-TH NUMBER
 80  FOR J=1 TO I—1
 90  IF N(I)>=N(J) THEN 200
 95  REM MAKE SPACE FOR THE I-TH NUMBER
100  X=N(I)
110  FOR H=I TO J+1 STEP —1
120  N(H)=N(H—1)
130  NEXT H
135  REM INSERT I-TH NUMBER IN POSITION J
140  N(J)=X
145  REM EXIT FROM J LOOP
150  J=I—1
200  NEXT J
210  NEXT I
215  REM PRINT SORTED NUMBERS
220  FOR I=1 TO K
230  PRINT N(I)
240  NEXT I
300  DATA 20:REM VALUE OF K
310  DATA 125,16,73,44,62,19,25,67,88,46
320  DATA 16,33,19,91,99,612,13,5,11,419
500  END
```

ENTER the program, and RUN it. (You needn't ENTER any of the REMarks. They are there just to help you follow the logic of the sorting algorithm.) When you finish, leave the program in the machine. We'll want to modify it in a minute.

Some features of the data become clearer once it has been sorted. For example, two numbers occur twice, namely, 16 and 19. A number that occurs with the highest frequency in the data set is called a "mode." In our case, there are two modes. (This data is "bimodal.") A mode is a kind of average. Another kind of average is the "median." It is the number in the middle of the (sorted) list, halfway between the top and bottom. Because there are an even number of terms in our list, two of them share the middle. In this case, the median is the ordinary ("mean") average of the two middle terms. For us, the median is (44 + 46)/2 = 45.

It turns out that statisticians are interested in many kinds of averages. This may come as a surprise; most of us are used to talking about THE average. Technically, the ordinary average (obtained by dividing the sum by the number of terms) is called the "mean," or "mean average."

Given K numbers, N(1), N(2), . . . , N(K), their mean can be computed by using the following:

```
400  A = 0
410  FOR  I = 1  TO  K
420  A = A + N ( I )
430  NEXT  I
440  PRINT  "THE  MEAN  IS" ; A / K
```

This routine may easily be added to the sorting program listed above.

Why so many averages? In a sense, an average represents a middle number. In the case of the mean, "middle" is understood in the sense of size. In the case of the median, "middle" is understood in the sense of position. And, in the case of mode, "middle" is realized in the sense of frequency. What is the best way to interpret "middle"? That depends. As we will see, different averages are best in different situations.

Suppose, for example, the XYZ Company employs 99 workers. Assume that each of these is paid $1000 a month while the owner is paid $1 million a month. Imagine that you are a newspaper reporter covering contract negotiations at XYZ. In an interview, the owner tells you the company employs 100 people at an average salary of $10,990 per month. (What average is he/she using?) In a separate interview, the chief labor negotiator claims that the average wage paid is only $1000 a month. (What average might he/she be using?) One can, perhaps, forgive each side's use of a self-serving figure, but which one do you think you'll use in your story? Which figure best represents the "middle"?

Suppose you are one of 99 heirs of a wealthy great-aunt. Upon her death, the estate consists of 99 $1000 bills and a bank account containing $1 million. Her will provides that the estate be averaged among her heirs, with one portion going to her attorney (as the executor). Rarely has a lawyer's job been easier. He simply distributes the median average to each blood heir, and keeps the odd "portion" for himself.

A well-chosen average can be very valuable in helping to "make sense" out of a lot of data. However, an average can never tell the whole story by itself. For example, 5 is the common value of the mean, median, and mode for each of these data sets.

A. 5, 5, 5
B. 4, 5, 5, 6
C. 1, 2, 4, 5, 5, 8, 10
D. 1, 5, 5, 5, 6, 6, 7

The "range" of the data is the difference between the largest and the smallest number. The ranges of the four data sets above are 0, 2, 9, and 6, respectively. A carefully chosen average, together with the size and range, provide an excellent beginning to the problem of understanding data.

Exercises (4.2)

1. Find the median and mode of each of the following data sets.

a. 10	b. 17	c. 12	d. 4	e. 9
5	44	17	16	1
15	6	12	8	5
1	53	31	2	
9	61	6	64	
5	8	9	32	
25	6	19		
8	2			
17				

2. What is the mean of the 20 numbers in the first paragraph of the section?

3. Find the mean, median, mode(s), and range of these 20 numbers,
   a. by hand
   b. by replacing the DATA statements in the program given in the text and using it

44	32	42	32	10
11	12	7	88	16
56	1	44	16	32
9	8	79	74	6

4. Give examples of data sets in which
   a. the median is smaller than the mean
   b. the mean is smaller than the median
   c. the mode is smaller than both the mean and the median
   d. the mode is larger than both the mean and the median

5. If you roll two dice 100 times and keep track of the total of each roll, what do you expect
   a. the mode to be?
   b. the range to be?

6. What is the probability that a randomly chosen letter in a textbook will be an "a"? a "b"? ... a "z"? While we have no way of calculating any of these probabilities, we can determine them empirically. Make a frequency distribution (See Section 3.1.) for the letters a–z, using 100 words from the middle of one of your texts.

7. Find two different data sets of seven numbers having the same mean, the same median, the same mode, and the same range.

8. Find a data set for which the mean and median are the same, but for which the mode is different from their common value.

9. Of the three averages discussed in this section, which is
   a. easiest to compute by hand?
   b. easiest to compute by computer?
   (Assume the numbers have not been sorted.)

10. Write a computer program to generate and PRINT 17 random numbers and to PRINT their mean.

11. Write a computer program to generate and PRINT 17 random numbers and to PRINT their median.

4.3 SQUARE ROOTS

What qualifies something to be an average? To simplify matters, let's suppose we're talking about a data set consisting of N positive numbers. Surely, an average should be a positive number, somewhere between the largest and smallest. A less obvious property is "homogeneity." If the numbers are $A(1)$, $A(2)$, ..., $A(N)$, and their "average" is A, then the "average" of the numbers $k*A(1)$, $k*A(2)$, ..., $k*A(N)$, should be $k*A$. One way to think of homogeneity is this: Suppose the data represents the heights of students in a high school class, measured in feet. Denote the "average" height by A. Then we should be able to obtain the "average" height in meters, for example, simply by converting A to meters. In this case, $k = .305$ (meters per foot). If we multiply each height by the conversion factor .305 and then average, we should obtain .305 times A, no matter what average we use.

Suppose a large Texas agribusiness were to go bankrupt. Assume the assets of the business consist of five square parcels of prime bottom land having respective dimensions 1, 1, 2, 5, and 7 miles on a side. The problem confronting the creditors is to ascertain the size of the average parcel. One of them sees it in his/her interest to argue for 2, the median. Another believes that the mean, 3.2, is the only *reasonable* average. Finally, the third finds that he/she would be better off if the mode, 1, were chosen as the representative parcel. Because they can't agree among themselves, they hire a consultant. After a minute, the consultant announces that the average parcel would be a square, 4 miles on a side.

Here is the consultant's reasoning: The value of property may depend on many things. But, the value of the agricultural property under consideration depends on its area. Thus the "average" parcel might best be described as one with the mean area. The sum of the area is

$$1 + 1 + 4 + 25 + 49 = 80$$

One fifth of 80 is 16. Thus, a square parcel of average area would measure 4 miles on a side.

The average used by the consultant is the "root-mean-square" (or RMS) average. This is a descriptive name. To obtain the RMS average of 1, 1, 2, 5, 7, we computed the mean of the squares of the numbers, and then took the "square root" of the result. The operation of taking

the square root is the opposite of the operation of squaring. One way
to indicate square root is by means of the symbol $\sqrt{}$. Here are a few
examples:

$$\sqrt{4} = 2; \qquad \sqrt{9} = 3; \qquad \sqrt{16} = 4; \qquad \sqrt{25} = 5; \qquad \sqrt{36} = 6$$

The difficulty arises with numbers like $\sqrt{2}$. The square root of 2 is a
number, x, such that $x^2 = 2$. What could x be? It can't be 1.5 because
$1.5^2 = 2.25$. So, 1.5 is too big. It can't be 1.4, because $1.4^2 = 1.96$. So,
1.4 is too small. (Remind you of Goldilocks choosing a chair?) But,
certainly 1.4 comes closer than 1.5. Maybe the answer is 1.41? No,
$1.41^2 = 1.9881$. What about 1.42? It's too big! It turns out that we
could continue in this way forever, thinking that the exact value is
right around the corner, that we're just about to find it. It isn't, and
we aren't. The square root of 2 is an "irrational" number with a decimal
expansion that never ends (and never repeats). We'll have more to say
about the theoretical aspects of irrational numbers in Section 6.8.† For
now, let's just talk about how to compute square roots (approximately).

BASIC comes with several built in "library" functions. The square
root is one of them. Try (successively) ENTERing these instructions.

```
PRINT SQR(49)
PRINT SQR(64)
PRINT SQR(81)
PRINT SQR(2)
PRINT SQR(-1)
```

Although the computer returns an answer for SQR(2), the answer
isn't exact. It's just the best the machine can produce without some
additional programming. (In fact, computers have never heard of irrational numbers.)

As you saw, the computer had problems with taking the square root
of a negative number. A moment's reflection shows why. The square
root of -1 is a number x, such that $x*x = -1$. Could x be positive?
No, a positive number times a positive number is always positive. Could
x be negative? No, a negative times a negative is positive. Maybe $x = 0$? It's the last hope (among the so-called "real" numbers), but it fails,
too. There is no "real" number whose square is -1.

---

†Suppose you pick a number, between 0 and 1, at random. You may be surprised to
learn that the probability of its being irrational is 1; but, this interesting discussion
will have to wait.

In reaching an understanding of why the computer has trouble with SQR(-1), we have inadvertently discovered another fact about square roots: Every positive number has two of them! Both 3 and -3 are square roots of 9. The computer assumes you know this and so outputs only the positive one.

Let's return to the RMS average. We can now give a formal definition. Suppose $A(1)$, $A(2)$, ..., $A(N)$ are N-positive numbers. Their RMS average is

$$\sqrt{\frac{A(1)^2 + A(2)^2 + \cdots + A(N)^2}{N}},$$

the Root of the Mean of the Squares. But, is the RMS average an average? It's pretty clear that the RMS of a set of positive numbers will be between the smallest and largest; but, is the RMS homogeneous? Yes. Here is the proof: For any positive number k,

$$\sqrt{([k*A(1)]^2 + [k*A(2)]^2 + \cdots + [k*A(N)]^2)/N}$$

$$= \sqrt{[k^2*A(1)^2 + k^2*A(2)^2 + \cdots + k^2*A(N)^2]/N}$$

$$= \sqrt{k^2*[A(1)^2 + A(2)^2 + \cdots + A(N)^2]/N}$$

$$= k*\sqrt{[A(1)^2 + A(2)^2 + \cdots + A(N)^2]/N}$$

Exercises (4.3)

1. Without using the computer, find the (positive) square root of
   a. 144   b. 256   c. 10,000   d. 625   e. 289
   (*Hint:* Make a guess, square it, and see if it works. If it doesn't, decide whether your guess was too big or too small before guessing again.)

2. Without using the computer, find the (positive) square root of
   a. 1.44   b. 0.01   c. 4.84   d. 5.29   e. 1.2321

3. Find
   a. SQR(5)   b. SQR(6)   c. SQR(7)   d. SQR(8)
   e. SQR(10)

4. Find
   a. SQR(.1)   b. SQR(.5)   c. SQR(1.23)   d. SQR(6.5)

5. Is SQR(1/X) = 1/SQR(X) for all positive X? Justify your answer.

6. Find the RMS average of 1, 2, 3, and 4.

7. Write a program to READ K, then READ K numbers and PRINT their RMS average.

8. Use the computer to find the RMS average of the 20 numbers in the first paragraph of Section 4.2. (*Hint:* Use Exercise 7.)

9. Use the computer to find the RMS average of the numbers in Exercise 3, Section 4.2.

10. Explain why the median and mode are homogeneous.

11. Prove that the mean is homogeneous.

12. The *curved* surface of a (right) cone has area $\pi r \sqrt{r^2 + h^2}$, where h is the altitude and r the radius of the base. Write a program to INPUT R and H, and to PRINT the *total* surface area.

13. Find the *total* surface area of a (right) cone if
    a. $r = 2$ meters and $h = 5$ meters
    b. $r = 2$ feet and $h = 3$ yards
    c. $r = 40$ cm and $h = 1.5$ meters

14. The equation $ax^2 + bx + c = 0$ can be solved by means of the *quadratic formula*:

$$x = [-b \pm \sqrt{b^2 - 4ac}]/2a$$

Write a program to INPUT A, B, and C, and to PRINT the solution(s) for x.

15. Find the solutions to
    a. $x^2 - x - 6 = 0$ b. $2x^2 + x = 5$
    c. $x^2 - x + 6 = 0$ d. $3x^2 + 12x + 1 = 0$
    e. $x^2 + 36 = 12x$ f. $x^2 = 2x$

4.4 STANDARD DEVIATION

Consider the distribution of numbers in a data set. It is our job to make some sense of the data, to achieve a quantitative understanding of the numbers as a whole. Step 1 is to compute a (well-chosen) average, to find the "middle." Step 2 is to determine how much the numbers vary, or deviate, from the average. The range gives a very rough picture of the variation. What we want to do now is develop a more sophisticated description of variability.

Suppose we have K numbers, $N(1), N(2), \ldots, N(K)$, whose mean is M. Consider one of the numbers. Call it N. (Maybe $N = N(5)$ or $N(17)$ or one of the other numbers.) The "deviation" of N from the mean is 0 if $N = M$; it is $N - M$ if $N > M$; and it is $M - N$ if $N < M$. This rather complicated definition can be simplified by making use of absolute value. Recall that $|X| = X$ if $X \geq 0$, and $|X| = -X$ if $X < 0$. Thus, the deviation of N from the mean is $|M - N|$.

Absolute Value is another of the BASIC library functions. It is denoted ABS. Experiment with it by ENTERing these instructions.

```
PRINT ABS(5)
PRINT ABS(-3)
PRINT ABS(0)
```

In a program, ABS($M-N(I)$) is the deviation of the I-th number from the mean.

Of course, one could define the deviation of each number from some other average. It may even be useful to do so. But, in the vast majority of real life applications, one is interested in deviations from the mean.

Having computed all the deviations, we have produced a new data set! We aren't going to do much with it, however. All we really want is its average. The mean average of the deviations is called (no surprises) the "mean deviation." The mean deviation gives us a feeling for how much the numbers (in the original distribution) vary from the mean, that is, a feeling for how much the numbers are spread out. The smaller the mean deviation, the closer the numbers are grouped around the mean. Let's do a couple of examples. In both of the distributions, A and B, which follow, the mean is 50.

A	Deviation	B	Deviation
44	6	16	34
45	5	34	16
45	5	45	5
47	3	45	5
50	0	50	0
52	2	52	2
53	3	55	5
54	4	61	11
54	4	66	16
56	6	76	26

Now,

$$[6 + 5 + 5 + 3 + \cdots + 4 + 6]/10 = 38/10$$

Thus, the mean deviation of distribution A is 3.8. Similarly, the mean deviation for distribution B is 12. "On the average," a typical number from data set A is 3.8 away from the mean, while a typical number from set B varies from the mean by 12. What we are doing is giving a quantitative description to what you can see at a glance: The numbers in set A are bunched close to 50, while the numbers in set B are much more spread out.[†]

It's interesting to note that while the numbers in set A may be "bunched close" to the mean, only four of them are within a single mean deviation (3.8) of the mean. On the other hand, in set B, where the numbers are "more spread out," six of them lie within a mean deviation of the mean, i.e., six of the numbers lie between 38 and 62.

What we would like is a theory which guarantees that a certain percentage of the data lies within one average deviation of the mean. Here, we face a little irony. If we insist on using the mean-average deviation, our theory will be awkward and difficult. If, however, we are willing to be more flexible in our choice of "average" for the deviations, a useful, elegant theory emerges.

[†] If this is obvious from glancing at the numbers, why go to all the trouble of computing the mean deviations? The point of these examples is to give you an intuitive understanding for the concept. Once that has been attained, you'll be ready to take on larger data sets, ones for which a glance will tell you nothing.

Although the reason is far from clear at this point, the "natural" average to use with deviations is the root-mean-square average! The RMS-average deviation is called the *standard deviation*.

By way of example, let's compute the standard deviation of the numbers in data set A.

Number	Deviation	Squared
44	6	36
45	5	25
45	5	25
47	3	9
50	0	0
52	2	4
53	3	9
54	4	16
54	4	16
56	6	36
		176

Now, $176/10 = 17.6$, and $\sqrt{17.6} = 4.2$. So, the standard deviation of these 10 numbers is 4.2. Similarly, the standard deviation of distribution B is 15.95.

For both of these examples, the standard deviation is greater than the mean deviation. In fact, it can be proved that

$$\text{RMS average} > \text{mean average}$$

of the same positive numbers, unless all the numbers are equal. So, if there is any variation at all, the standard deviation is always larger than the mean deviation!

The Russian mathematician P. L. Chebychev[†] (1821–1894) discovered the following remarkable result: For any distribution of K numbers, and for any $h > 1$, at least $[1 - (1/h^2)]*K$ of the numbers lie within h standard deviations of the mean. Here are a few examples for various values of h.

[†] Also translated "Chebichev," "Chebyshev," or "Tchebycheff."

h	$1-(1/h^2)$
2	.75
3	.89
4	.94
5	.96
6	.97

In data set A, for example, two standard deviations are 8.4. Since K = 10, this means that at least eight of the numbers must be in the range 41.6 to 58.4. In fact, all the numbers are in this range. For distribution B, two standard deviations are 31.9. So, Chebychev's Theorem guarantees that at least eight of the numbers will be in the range 18.1 to 81.9. In fact, nine of the numbers lie in this range.

Exercises (4.4)

1. If $|x-5| < 1$, show that $4 < x < 6$.
2. If $ABS(X-2) < 3$, show that $-1 < X < 5$.
3. If $|x-3| < 2$, show that $1 < x < 5$.
4. By hand, without using a computer, find the mean deviation of each of the following data sets. (*Hint:* First calculate the mean.)
   a. 1, 6, 11, 16, 21
   b. 2, 4, 8, 16, 32, 64
   c. 3, 9, 27, 81
   d. 1, 1, 2, 3, 5, 8, 13, 21, 34, 55, 89
5. Write a computer program to INPUT (or, if you prefer, READ)
   a. K
   b. K non-negative numbers
   and output
   c. their mean
   d. their mean deviation
6. Compute the mean and mean deviation for each of these data sets. (*Hint:* Make use of your solution to Exercise 5.)
   a. 16, 9, 21, 14, 62, 31, 5, 9, 10, 2
   b. 11, 13, 8, 60, 12, 4, 19, 11, 6, 14
   c. 8, 18, 76, 55, 64, 3, 23, 43, 1, 5

7. By hand, without using a computer, compute the mean and standard deviation of 5, 8, 9, 11, 17.

8. Write a computer program to INPUT a data set of at most 1000 positive numbers (without having to INPUT or READ the size of the set), and to output the mean and standard deviation of the set.

9. Use the program you wrote for Exercise 8 to compute the mean and standard deviation of each of the data sets in Exercise 6.

10. Confirm, for each of the data sets in Exercise 6, that the mean deviation is less than the standard deviation.

11. Confirm Chebychev's Theorem for each of the data sets in Exercise 6.

4.5 NORMAL DISTRIBUTIONS

In this section we wish to consider, conceptually at least, very large sets of data. One example of this kind of situation arises in nationally administered college entrance tests in which tens of thousands of individuals take the same test. By itself, an individual score is pretty useless. More meaningful is some sort of "standardized score" showing how a particular performance compares with all the others. Standardized scores are based on an analysis of the frequency distribution of all the scores.

Recall Chebychev's Theorem which asserts that for any data set of K numbers, and for any $h > 1$, *at least* $[1 - (1/h^2)]*K$ of the numbers lie within h standard deviations of the mean. It is results of this type that allow us to do the kind of analysis we need to understand not only college entrance tests, but many other sets of data as well.

It is sometimes possible to do better than Chebychev's Theorem for certain special kinds of data.[†] One of these involves the "bell-shaped" curve in Figure 4.1. It illustrates what has come to be known as a "normal" frequency distribution. A precise definition of the normal

---

[†]Chebychev's Theorem works for *any* data set. We can "do better," but only at the cost of limiting the scope of our discussions to a certain restricted kind of data.

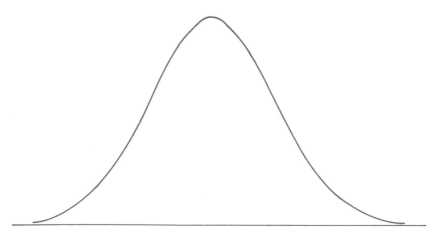

Figure 4.1

distribution depends on some ideas and techniques from Calculus; but, we can base an intuitive discussion on the binomial distribution.

Consider the situation in which n coins are tossed simultaneously. As we know, the probability that exactly r heads will turn up is $nCr/2^n$. Suppose we toss the n coins not once, but repeatedly, a total of 2^n times. Then the expected number of times exactly r heads will occur is nCr. In an (idealized) frequency distribution involving 2^n tosses of n coins, exactly r heads will occur nCr times. To make things more definite, let's take n $=$ 10. Then 2^{10} $=$ 1024; there are 1024 possible outcomes, and we plan for the 10 coins to be tossed 1024 times. If, for example, r were six, then the expected frequency of six heads (and four tails) is $10C6$ $=$ 210.

Figure 4.2 shows a graph of r $vs.$ $10Cr$. This is an idealized frequency distribution graph, one in which every outcome occurs with its expected frequency. The graph illustrates the binomial distribution.[+] It's just as if we were to examine a data set consisting of 1024 numbers, each of which comes from 0–10. In this set, 6 occurs $10C6$ $=$ 210 times, 5 occurs $10C5$ $=$ 252 times, and, more generally, r occurs $10Cr$ times.

As a first step in analyzing this distribution, observe that the mean of the collection of 1024 numbers is 5. (You don't have to calculate it!

[+] Strictly speaking, this is the binomial distribution for $p = 1/2$, i.e., when both outcomes (heads and tails) are equally likely. (See Section 3.10.)

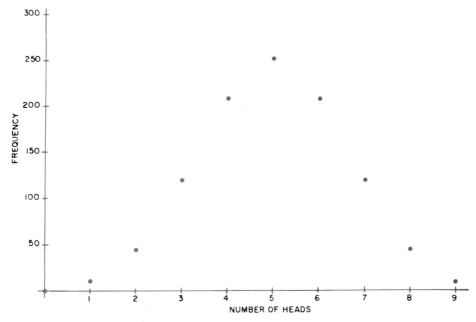

Figure 4.2

Just notice that the 210 4's balance the 210 6's, that the 120 3's balance
the 120 7's etc.) Indeed, the mean of an ideal binomial distribution is
always $n/2$. (Moreover, the mean, median, and mode are all the same.)

Another nice fact about these distributions is that the standard deviation can be calculated with ease. It turns out always to be $\sqrt{n}/2$.
Let's confirm that for the case at hand, namely $n = 10$.

Deviation	Squared	Frequency	Contributes
5	25	1	25
4	16	10	160
3	9	45	405
2	4	120	480
1	1	210	210
0	0	252	0
1	1	210	210
2	4	120	480
3	9	45	405
4	16	10	160
5	25	1	25
			2560

Now, $\sqrt{(2560/1024)} = \sqrt{10}/2$.

Keeping Figure 4.1 in mind, let's connect the dots in Figure 4.2 with a smooth curve. The result is shown in Figure 4.3. Thus, using a lot of imagination, we are able to pass from the ideal binomial distribution in Figure 4.2 to something very like Figure 4.1. Needless to say, this imaginative leap requires further justification, and this is where we need Calculus. The crucial fact turns out to be this: As we take n larger and larger, the binomial distribution gets closer and closer to the normal distribution. The "limiting case" of the binomial distribution "as n goes to infinity" is the normal distribution.

Before going on, it should be pointed out explicitly that the binomial curve in Figure 4.3 can be made to appear in a somewhat different guise simply by changing the scales of the horizontal and vertical "axes." (See Figure 4.4.) It is the general, qualitative bell-shape of the curves in Figures 4.1, 4.3, and 4.4 that interests us here.

The next question is whether the normal distribution is any good in real life situations. The surprising answer is a resounding YES! If you took a large, random sample of people and measured their heights, weighed them, counted the hairs on their heads, or administered a standardized test, you would find the results in very good agreement

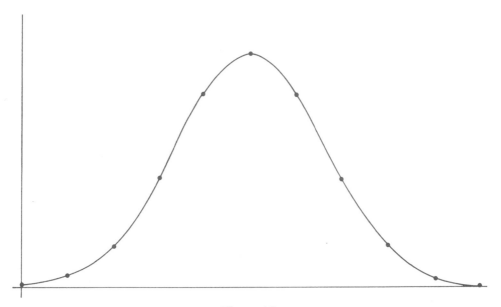

Figure 4.3

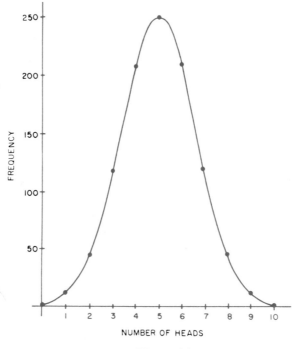

Figure 4.4

with the normal distribution. The numbers of stars in different sectors of the sky, the number of grains of sand in a jillion pint containers, the volumes of all the soup bowls in the world, and so on. The normal distribution models a great many real life statistical situations.

The normal distribution is so useful, it has received extensive study over many years.[†] One result of all this attention is a variation on Chebychev's Theorem. In a *normal* distribution, 68% of the data is within one standard deviation of the mean, 95% within two standard deviations, and 99% within three. (Compare this with Chebychev's guarantee of at least 75% within two standard deviations and 89%

[†] The normal distribution seems first to have been studied by Abraham de Moivre (1667–1754). It was fully developed by Carl Friedrich Gauss (1777–1855) and Pierre S. LaPlace (1749–1827). Applications to biological data were established by Sir Francis Galton (1822–1911). The normal distribution is sometimes referred to as the Gaussian distribution.

within three, with no information on one standard deviation.) The new result is illustrated in Figure 4.5.

It needs to be emphasized that the percentages illustrated in Figure 4.5 are for normal (frequency) distributions. If a data set is only approximately normally distributed, then the percentages are correspondingly approximate. Moreover, whereas Chebychev's Theorem guarantees percentages of "at least" so and so, the percentages given by this new result may be in error in either direction when the distribution is not normal.

Let's take a practical example. Suppose the Department of Defense were to buy 100,000 light bulbs. Tests reveal that the bulbs can be expected to burn a (mean) average of 100 hours with a standard deviation of 8 hours. This means that 68% of the bulbs will have a life of 92 to 108 hours. That's 68,000 bulbs! On the other hand, only 500 bulbs or so (one-half of 1%) will still be burning after 124 hours (three standard deviations above the mean).

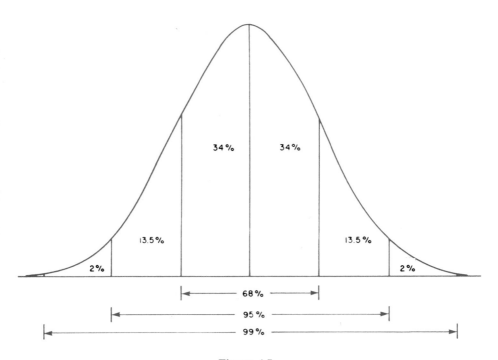

Figure 4.5

Exercises (4.5)

1. Grades on a test tend to be (approximately) normally distributed provided the test is neither too easy nor too hard, and provided enough people take the test. Thus, when students ask to be graded "on the curve," they are asking that any score within 1 standard deviation of the mean correspond to a C, that A's be reserved for those at least 2 standard deviations above the mean, that D's be given to all those between 1 and 2 standard deviations below the mean, etc. Write an essay discussing the pros and cons of "grading on the curve."

2. Suppose that people's heights, measured in inches, are normally distributed with a mean of 66 and a standard deviation of 3.[†] In a city of 1 million people, how many would you expect to
   a. be taller than 6 feet?
   b. be between 5 and 6 feet tall?

3. The life span of a certain type of water heater is approximately normally distributed with a mean of five years and a standard deviation of two years. If the company guarantees the product for one year, what fraction of the water heaters will it have to replace?

4. One fact of life in American education is the standardized test. Suppose you were to take one of these tests. If your numerical score is s, the mean is m, and the standard deviation is d, then your "standardized score" is $(s - m)/d$. The effect of this standardization is to report your score in terms of the number (or fraction) of standard deviations above or below the mean. Determine your standardized score if
   a. $s = 64$, $m = 50$, $d = 20$ b. $s = 40$, $m = 50$, $d = 20$
   c. $s = 80$, $m = 74$, $d = 10$ d. $s = 85$, $m = 60$, $d = 18$

5. Rank the performances on the four tests in Exercise 4, from worst to best.

6. Suppose you have three big final exams: History, English, and Mathematics. Suppose your scores are given in the table below.

[†] The tallest person who ever lived seems to have been Robert Wadlow of Alton, Illinois. Wadlow was 8 feet, 11 inches tall.

On which test did you perform the best, relative to the other students? On which the worst? (*Hint:* See Exercise 4.)

Subject	Score	Mean	s.d.
English	78	65	10
History	85	67	12
Math	61	50	5

7. If grades are assigned "on the curve" in each of the classes in Exercise 6, what grades would you receive? (*Hint:* See Exercise 1.)

8. If the IQ's of elementary school pupils are measured by a certain test having a mean of 100 and a standard deviation of 15, what is the probability that a randomly chosen child will have an IQ above 130?[†]

9. Write a program to INPUT or READ a data set and to output the percentage of the data lying within 1 standard deviation of the mean.

10. Are the random numbers produced by RND normally distributed? If so, 68% of them should lie within a standard deviation of the mean. Modify the program you wrote for Exercise 9 so that it will internally produce a data set of 100 RaNDom numbers, and determine what percentage of them lie within 1 standard deviation of the mean. RUN the program several times and report the results.

11. If the standard deviation of a set of numbers is 0 (zero), prove that all the numbers must be the same.

12. Imagine a College Admission Test with a mean of 500 and a standard deviation of 100. Suppose the competing University Aptitude Exam affords a mean of 30 and a standard deviation of 10. If Walter Edmonds College considers the tests to be comparable, requiring each applicant to take one or the other, who will impress the admissions committee more, Joan with a CAT score of 645, or Paul with a UAE score of 42?

[†] It has been said that a little knowledge can be a dangerous thing. We have just scratched the surface of the statistics used to analyze standardized tests. Another important statistic in this regard is the "standard error" which gives the probability that the reported standardized score is accurate, say, to within 5 points.

13. Suppose that a large department store is about to order a supply
of men's shirts for the spring. Assume that men's neck sizes average 15½ with a standard deviation of one-half size. How many
shirts of each neck half-size (say in the range of sizes 14–17) should
be ordered per 1000 shirts? (*Hint:* Obviously, the store doesn't need
the answer to two decimal places! What's wanted is a "ball park
estimate." Take advantage of the fact that the binomial distribution approximates the normal distribution. Since 2^{10} is about
1000, use $n = 10$. Make use of Figure 4.4 to read off appropriate
values. To make the sum come out right, these "appropriate values" will have to be rescaled and "fudged."[†])

4.6   CHI-SQUARED

In a normal distribution, roughly 34% of the data lies between the
mean and the first standard deviation above the mean; roughly 47.5%
lies between the mean and two standard deviations above the mean,
and so on. (See Figure 4.5, Section 4.5.) But, what about half a standard
deviation or 2.1 standard deviations? The following table gives the
fraction of the data between the mean and s standard deviations above
(or below) the mean for s running from 0.1 to 3.0 in steps of one-tenth.

Suppose you were to receive the results of a standardized test. If you
were in the 81st percentile, what would it mean? Since the 50th percentile corresponds to the mean, the 81st is 31% above half. Looking
at the % column in Table 4.1, 31 lies somewhere between 0.8 and 0.9
standard deviations above the mean.

Our next topic is a little more complicated. It involves the so-called
chi-squared statistic, C^2 for short. C^2 was devised about the turn of
the century by Karl Pearson (1857–1936) to address questions like the
following: How does one determine whether a die is fair?

Throughout our discussion of probability in the previous chapter, we
made certain assumptions based on "equal likelihood." A "fair" coin,

---

[†] Here, again, our job would be made easier if we had the techniques of Calculus at
our disposal.

Table 4.1 Standard Deviations

s	%	s	%
0.1	4.0	1.6	44.5
0.2	7.9	1.7	45.5
0.3	11.8	1.8	46.4
0.4	15.5	1.9	47.1
0.5	19.1	2.0	47.7
0.6	22.6	2.1	48.2
0.7	25.8	2.2	48.6
0.8	28.8	2.3	48.9
0.9	31.6	2.4	49.2
1.0	34.1	2.5	49.4
1.1	36.4	2.6	49.5
1.2	38.5	2.7	49.6
1.3	40.3	2.8	49.7
1.4	41.9	2.9	49.8
1.5	43.3	3.0	49.9

for example, is one that is equally likely to come up heads or tails. A fair die is one for which each face is equally likely to come out on top when the die is rolled in a random fashion. Let's come straight to the hard part. How does one tell whether a particular die is fair? Test it? Suppose your test consisted of rolling the die 60 times and recording the results. Of course, the expected number of 1's would be 10. The expected number of 2's, 3's, 4's, 5's, and 6's would also be 10. But, we know enough about probability theory to be very surprised if each number were to occur exactly 10 times. What if the recorded frequencies turned out to be these:

Value	1	2	3	4	5	6
Observed frequency	16	6	7	7	16	8
Expected frequency	10	10	10	10	10	10

It seems there are far too many 1's and 5's, while every other number is under-represented. How unlikely are these outcomes if the die were fair? Put another way, what is the probability of observing discrepancies this large (or bigger!) in a fair die? This is the question that the C^2 statistic answers.

The computation of C^2 is not unlike the computation of the standard deviation. For each value, square the deviation from expectation, and divide by the expected frequency. When you've done that for each value, sum up your individual answers. That's the C^2 statistic. In our case, the computations are:

$$(-6)^2/10 + (4)^2/10 + (3)^2/10 + (3)^2/10 + (-6)^2/10 + (2)^2/10$$

$$= (1/10)*(36 + 16 + 9 + 9 + 36 + 4) = 110/10 = 11$$

So, in our example, chi-squared is 11. When the observed frequencies are very different from the expected frequencies, C^2 is large. Indeed, C^2 is zero only if every single observed frequency is exactly equal to its expected frequency. But, what does a C^2 value of 11 tell us? The answer to that lies in Table 4.2, and we are almost ready to use it. There is just one more sticky detail to clear up. It involves the first column of Table 4.2 that is (somewhat cryptically) headed "DEG." We just observed that C^2 is zero only if all the observed frequencies are exactly equal to the corresponding expected frequencies. Could all but one of the observed frequencies have been exactly what was "expected" with only one showing a deviation? Of course not. Then the numbers wouldn't add up to the correct total. In fact, it was really unnecessary to report the observed number of 6's. Since 16 + 6 + 7 + 7 + 16 = 52, and the die was rolled 60 times, there must have been eight 6's. This explains what a statistician means by "degrees of freedom," or "DEG" for short. The last observed frequency is not "free" since it can be determined from the others. The number of degrees of freedom is one less than the number of possible outcomes; in our case, DEG = 5.

Now, we're ready to use Table 4.2. What one does is look along row 5 for the computed C^2, namely 11. Finding 11 (in the 6th column), one

Table 4.2 Chi-squared (C^2) Statistic

DEG	.99	.95	.9	.5	.1	.05	.01
1	.0002	.004	.02	.46	2.7	3.8	6.6
2	.02	.10	.21	1.4	4.6	6.0	9.2
3	.12	.35	.58	2.4	6.3	7.8	11
4	.30	.71	1.1	3.4	7.8	9.5	13
5	.55	1.1	1.6	4.4	9.2	11	15
6	.87	1.6	2.2	5.3	11	13	17

finds a probability, .05, at the *top* of that column. This probability, .05, is the chance that a fair die would result in discrepancies as large as or larger than those observed. We can loosely interpret our finding this way: There is one chance in 20 that the die involved in the experiment is a fair one.

It is a little frustrating not to be certain. The die might very well be fair, but would you want to bet that it is?[†]

What about another example? Suppose we tossed a coin 60 times and produced 39 heads (and 21 tails). Then

$$C^2 = (30-39)^2/30 + (30-21)^2/30$$

$$= (1/30)*(81+81)$$

$$= 5.4$$

Since there are two possible outcomes, DEG = 1. Looking along the first row of Table 4.2, we encounter one of its drawbacks: 5.4 is nowhere to be seen. The table is incomplete.[‡] If it were complete, we would expect to find 5.4 somewhere between 3.8 and 6.6. Thus, the probability we're looking for must be between .05 and .01. So, the probability that a fair coin would produce 39 (or more) heads in 60 tosses is somewhere between 1/20 and 1/100.

There is another way to look at the chi-squared table. Suppose a student is given the following homework assignment: Over the weekend, roll a die 600 times and turn in the resulting frequency distribution. Here is what the student handed in on Monday morning:

Value	1	2	3	4	5	6
Observed frequency	99	103	99	101	98	100
Expected frequency	100	100	100	100	100	100

[†]We do not always extend to others the rights we claim for ourselves. In particular, the die need not be considered "fair" until proved otherwise. (Even a fair die might produce sixty 1's in a row!)

[‡]The Chemical Rubber Company (for one) publishes a handbook of tables and formulas from which a more complete table can be obtained. It is called "C.R.C. Standard Mathematical Tables."

The instructor, more than a little suspicious that the observed frequencies should be so close to the expected ones, decides to subject the
data to a chi-squared test. The computation is

$$C^2 = (1 + 9 + 1 + 1 + 4 + 0)/100$$

$$= 16/100 = .16$$

Checking row 5 of Table 4.2, we find that .16 is well off the chart.
Here is what the instructor tells the student: "It seems the probability
that a fair die would lead to discrepancies as large or larger than the
ones you reported is on the order of .999. That's the good news. The
bad news is that the probability a fair die would produce discrepancies
no bigger than this is on the order of $1 - .999 = .001$. There is maybe
1 chance in a thousand that your data was produced in an honest
experiment."

Exercises (4.6)

1. Suppose the results of a certain test are normally distributed with
   a mean m and a standard deviation d. It is usual to convert each
   numerical score s to a *standardized score* $z = (s-m)/d$. Thus, s
   represents z standard deviations from the mean. Use Table 4.1 to
   find the percentile ranking if
   a. $m = 35$, $d = 5$, $s = 42$ b. $m = 50$, $d = 10$, $s = 71$
   c. $m = 100$, $d = 10$, $s = 101$ d. $m = 100$, $d = 25$, $s = 80$
   e. $m = 52$, $d = 10$, $s = 35$ f. $m = 67$, $d = 13$, $s = 67$

2. Suppose that a certain brand of washing machine has a mean life
   span of 14 years, with a standard deviation of 4 years. What fraction
   of the machines sold 20 years ago are still running?

3. Suppose that the heights of men, ages 20–30, are normally distributed with a mean of 70 inches. In a city with 100,000 men in the
   20–30 age bracket, how many will stand taller than 6 feet 4 inches
   if the standard deviation is
   a. 4 inches?
   b. 3 inches?

4. Suppose the mean family income in a certain state is $18,000, with
   a standard deviation of $5000. The State Legislature decides to raise
   the official "poverty line" from $8000 to $9000. If the state contains

1 million families, how many more of them will become "poor" as a result of this new definition of poverty?

5. Write a program to INPUT or READ a data set and to output the percentage of the data lying within H standard deviations of the mean, where H is a number to be INPUT when the program is RUN. (*Hint:* See Exercise 9, Section 4.5.)

6. Suppose you wanted to do a chi-squared test on a frequency distribution from a 12-sided die. How many degrees of freedom would there be?

7. Let's see how "fair" (or, maybe, "random"?) the computer's RaNDom number generator is. Program the machine to stimulate the rolling of a single die 600 times and to output its "observed frequency" distribution. Compute the resulting C^2 statistic, and find the (approximate) probability that a "fair" machine would produce discrepancies as bad or worse than those produced by your machine. Here is some more of the C^2 table, for DEG = 5.

PROBABILITY	.8	.7	.3	.2
C^2	2.3	3.0	6.1	7.3

8. RUN your program from Exercise 7 several (more) times, and record the results.

9. Write a program (part of which might be suitable for use as a subroutine in another program) to READ degrees of freedom, observed frequencies and expected frequencies, and to output the chi-squared statistic.

Chapter 5
NUMBER BASES

5.1 NUMERALS

A "numeral" is a name for a number. In the old Roman system, "IV" was the name given to four. We commonly speak of IV as "Roman numeral" four. The ancient Hindu symbol for four was " Y ," while the Arabic numeral is " C ." The difference between a number and a numeral is rather like the difference between you and your name, or between having a $100 bill and knowing what to call one when you see it. Of course, our numerals—a modification of Arabic numerals—are

$$0, 1, 2, 3, 4, 5, 6, 7, 8, 9, 10, 11, 12, \ldots$$

The Roman numerals were designed to make recognition easy. Our present day numerals have the advantage of simplifying computation. (Can you imagine trying to multiply MCMXLIX and MMCMLXXVIII entirely within the Roman numeral system?)

Let's take a good look at our numerals. We have ten different symbols (or digits) which can be used, either singly or in groups, to represent numbers. When grouped, position becomes important. For example, 82 and 28 represent different numbers. They are different numerals.[†]

[†] In the ancient Egyptian Hieroglyphic numeration system, position was not important. A "lotus flower" ($\overline{\lambda}$) always represented one thousand, regardless of its position relative to any other "digits."

After inventing individual symbols for the first few numbers, a difficulty arises. It is clear that one cannot go on inventing new, original symbols for each successive number indefinitely. Eventually, it becomes necessary to combine the symbols already available in order to produce new names. In our case, that time comes after the tenth symbol. (Starting from 0, 9 is the tenth numeral.) Why do you suppose there are precisely ten symbols and not more or fewer?

It is no accident that our fingers (and toes) are known to biologists as "digits." It seems clear enough that early peoples used their fingers (and maybe even toes) as computational aids.[†] That we have exactly ten individual numeration symbols is a consequence of the fact that we have ten fingers.

How would things have been different had we evolved with four fingers on each hand? What if intelligence had "gone to the birds" with three fingers on each "hand"? It is sometimes interesting to speculate about life on other planets. How would the fact that Vegans have four fingers on each hand affect their mathematics (and, therefore, their computer science, their physics, . . ., even their bank accounts)? Let's see!

Suppose we assume that the Vegans are like us in all ways except for having four fingers on each hand. Being like us, they would have developed a numeration system like ours, starting off, like ours, with the numerals 0, 1, ... But, unlike us, they would have stopped with the eighth numeral, namely, 7.

How would the Vegans denote eight (the next number after 7)? They would do what we have done and start using position. To the Vegan's, "10" is the numeral for eight. After 10 comes 11, just as it does with us, but "11" is the Vegan numeral for nine, not eleven!

Suppose a Vegan wrote "35." What number would he/she have in mind? To answer this question, we need to examine a little more carefully what we mean by "35." The "3" is in the "tens" position, meaning that it represents three tens. The "5" is in the "ones" position, so it represents five ones. Three tens and five ones is the number (not the numeral) thirty-five. What about the Vegan? To him/her, the "3" is in the eights position, meaning that it represents three eights! The "5"

[†] How many kindergarten children use their fingers to help them with addition problems?

is in the "ones" position, even for a Vegan. Thus, to a Vegan, "35" is the numeral for three eights and five ones, that is, for twenty-four plus five, or twenty-nine.

The last thing most of us want is to be laughed at. If it were to leak out that we're talking seriously about "Vegans" . . . well, let's not take any chances. It's better to use the dry, formal language of dusty old books than to be made fun of. The formal words for the "Vegan" system are "base eight." By way of contrast, our system is known as "base ten." What we are going to do is spend some time learning about base eight. Let's, in fact, agree now to abbreviate this as "base 8." (The "8" in "base 8" is the base ten numeral for eight. This shouldn't cause any confusion because the Vegans, together with all base 8 people, have never heard of "8." Remember, their largest single numeral is "7.")

Let's become more familiar with converting (or translating) base 8 numerals to base 10 numerals. What about 123? The 3 is in the ones position, or the ones "column"; it has the same "value" in both systems. The 2, lying in the eights column, has the value sixteen (two eights). What about the 1? Is it in the eighties column? In our system, we go from the tens column to the hundreds, and then to the thousands, . . . , multiplying by ten at each stage. What is the comparable procedure in base 8? We should multiply by eight at each stage! If we think of our base 10 "columns" as being labeled by powers of ten, the base 8 columns will be labeled, similarly, by powers of eight. After the eights column, comes the eight-squared column. The "1" in 123 is in the sixty-fours column. Thus, 123(base 8) = 83(base 10) because (using base 10 numerals) $1*64 + 2*8 + 3*1 = 83$.

What is the next column to the left of the sixty-fours column in base 8? It is $8*64 = 8^3 = 512$. Thus, to convert 3210(base 8) to base 10, we compute

$$3*512 + 2*64 + 1*8 + 0*1 = 1536 + 128 + 8 \qquad (5.1)$$
$$= 1672,$$

that is, 3210 is the base 8 name for the number one thousand, six hundred, seventy-two. It is important to emphasize that the *number* is the same in both systems. We write it as "1672," while the Vegans write it as "3210." If we count the number of people on a Vegan space ship and obtain 1672, their name for the same number of people is 3210. (The crew of a space ship doesn't double just because its people have four fingers on each hand.)

On his "Voyage to the Country of the Houyhnhnms,"[†] Lemuel Gulliver discovered an interesting place in which intelligence resided in creatures that we would identify as horses. (The human-looking animals, the Yahoos, were totally uncivilized, to say the least.) Naturally, having a single "digit" on each "hand," the Houyhnhnms might be expected to conduct business in base 2. Consider, for example, the base 2 numeral 1,010,011.[‡] Let's convert it to a base 10 numeral.

Proceeding from right to left, we come to the 1's column, the 2's column, the 2-squared or 4's column (where we encounter our first zero), the 2-cubed or 8's column, the 16's, 32's and finally the 64's column. This prodigious (looking) number is

$$1*64 + 0*32 + 1*16 + 0*8 + 0*4 + 1*2 + 1*1 = 83,$$

that is, 1,010,011(base 2) = 83(base 10).

In base 10, there are ten individual symbols, 0–9. In base 8, there are eight, 0–7. In base 2, there are only two, 0 and 1! Indeed, the first few counting numbers in base 2 are 1, 10, 11, 100, 101, 110, 111 (equals seven), etc.

Of course, the land of the Houyhnhnms is fanciful. How fanciful are base 2 numbers? Probably, you wouldn't like to do any extended computations in base 2. It may be a surprise to learn, therefore, that your computer does all of its arithmetic (and everything else!) in base 2.

One final definition: In base 8, we spoke of the 1's column, the 8's column, the $8^2 = 64$'s column, and so on. It turns out to be convenient to define $8^1 = 8$ and $8^0 = 1$. The same thing goes for 2 or any other number. Specifically,

$$r^1 = r \uparrow 1 = r, \text{ for any } r \text{ while}$$
$$r^0 = r \uparrow 0 = 1, \text{ for any } r \neq 0.$$

Confirm that your computer understands these conventions by ENTERing

```
PRINT 5↑1
PRINT 5.63↑1
PRINT 7↑0
PRINT 3.14159↑0
```

[†] Pronounced "Whinnims." See *Gulliver's Travels* by Jonathan Swift.

[‡] Commas are typically omitted from base 2 numerals in the computer science literature. The problem is that a computer would interpret 1,010,011 as three numbers.

Exercises   (5.1)

1. Write the first fifteen counting numbers (one through fifteen) in each of the following number bases:
   a. base 8   b. base 2   c. base 5   d. base 6

2. Convert each of the following base 8 numerals to a base 10 numeral:
   a. 707   b. 1011   c. 747   d. 737   e. 12,345

3. Convert each of the following base 2 numerals to a base 10 numeral:
   a. 111   b. 1000   c. 111,111   d. 10   e. 10,101

4. Express the next number, after the given number, in the same number base:
   a.   555(base 6)   b. 707(base 8)   c. 1011(base 2)
   d. 1011(base 3)   e.1234(base 5)   f.   33(base 4)

5. Write a computer program to convert base 8 numerals to base 10 numerals.

6. Write a computer program to convert base 2 numerals to base 10 numerals.

7. Write a computer program to INPUT B and then convert a base B numeral to a base 10 numeral.

8. Consider the 8 rows of Figure 5.1 separately. In each row, interpret a dark square as a 1 and a light square as a 0. Use this interpretation to convert each row to a base 2 numeral. Finally, convert

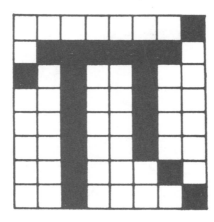

Figure 5.1

each base 2 numeral to a base 10 numeral. Record the sequence of (eight) base 10 numerals. (See the telemetry discussion at the end of Section 3.10.)

9. Try to reverse the procedure of the previous problem to convert the following sequence of base 10 numerals into a picture on an 8-by-8 grid: 20, 42, 81, 81, 81, 50, 28, 16.

10. Confirm that the computer obeys the usual laws of exponents in the following instances.

   a. $(2 \uparrow 3) * 2 \uparrow (-3) = 2 \uparrow 0$ b. $(2 \uparrow 5) / (2 \uparrow 4) = 2 \uparrow 1$
   c. $(5 \uparrow 4) * 5 \uparrow (-4) = 5 \uparrow 0$ d. $1 / (2 \uparrow (-1)) = 2 \uparrow 1$

11. Which sets of parentheses are actually needed in Exercise 10?

5.2 ARITHMETIC

In the last section, we discussed conversion from base 8 to base 10. What about going the other way, that is, converting from base 10 to base 8? Let's try a nice round number, say 2001. What we want to do is reverse the procedure of the previous section. (See Equation (5.1), in particular.)

Here's an idea. Let's first determine the left-most column that will be needed in the representation of 2001 in base 8. Reading from right to left, the base 8 column headings are 1's, 8's, 64's, 512's, and so on, the next one being $8*512 = 8^4 = 4096$. Since 4096>2001, we won't be needing any 4096's in our base 8 representation of 2001.

Pick the largest (base 8) column heading less than 2001. It's 512. There are three 512's in 2001 with a remainder of 465. Thus, the base 8 representation of 2001 is $3XYZ$, where X, Y, and Z are yet to be determined. But, XYZ must just be the base 8 representation of 465. (We have already "taken care" of $3*512 = 1536$ out of the 2001. Only 465 more remains to be "distributed" among the "lower" base 8 columns.)

To determine X, we ask how many times 64 goes into 465 (the remainder of the previous division problem). The answer is 7 with a remainder of 17. So, $X = 7$. (If X had turned out to be 8 or more, we would know we'd made a mistake. Why?)

So far we have 37YZ(base 8) = 2001(base 10) and we want to determine Y and Z. But, they make up what is left over after using up $3*512 + 7*64$ of the 2001; that is, YZ is the base 8 representation of the remainder, 17. We now ask how many 8's there are in 17. The answer, of course, is 2 with a remainder of 1. Thus, $Y = 2$, and we have just 1 left over. The ones digit, Z, must therefore be 1. So,

$$2001\text{(base 10)} = 3721 \text{ (base 8)}.$$

Let's call the approach just used Algorithm I. Looking ahead to writing a BASIC program to do the conversion, we need to ask ourselves how we're going to explain this algorithm to the computer; that is, how we're going to write the program. We'll have to INPUT a base 10 numeral, decide how many digits will occur in the base 8 numeral, and then start dividing, being sure to keep track of the quotients and remainders. Suppose we give it a try.

```
10 INPUT "WHAT IS THE BASE 10 NUMERAL";N
```

So far, so good. Now, we need to know the highest power of 8 less than N. Let's call that P.

```
20  IF N>7 THEN 30
25  PRINT "IN BASE 8 IT'S ":GO TO 200
30  P=1
40  P=P+1
50  IF 8↑P<=N THEN 40
60  P=P−1
```

Now we're ready to start dividing.

```
70   FOR I=P TO 1 STEP −1
80   Q(I)=INT(N/8↑I)
90   R(I)=N−Q(I)*8↑I
100  N=R(I)
110  NEXT I
```

It remains to assemble and PRINT the answer.

```
120  PRINT "THE BASE 8 NUMERAL IS"
130  FOR I=P TO 1 STEP −1
140  PRINT Q(I);
150  NEXT I
160  PRINT R(1)
200  END
```

(Naturally, if we expect to deal with numbers whose base 8 numeral has more than ten digits, DIMension statements will have to be added.) RUN the program (at least) for N = 2001.

While Algorithm I may be natural from our perspective, it is not the best algorithm available from the computational standpoint. (The little subroutine in lines 40–60 can be avoided.) Here is another approach, call it Algorithm II. Let the base 10 numeral be N. Divide N by 8, obtaining a quotient Q1 and a remainder R1. (In Figure 5.2, Q1 = 250 and R1 = 1.) Then R1 is the ones digit in the base 8 numeral. (Have another look at Equation (5.1).)

Divide Q1 by 8, obtaining a quotient Q2 (= 31 in Figure 5.2) and a remainder R2 (= 2 in Figure 5.2). Then R2 is the eights digit in the base 8 numeral. Divide Q2 by 8 to get a quotient Q3 (= 3 in our example) and a remainder R3 (= 7 in Figure 5.2). Then R3 is the 64's digit. Continue in this way until you reach a quotient Qk (in the example, k = 4), which is zero. Then Rk (R4 = 3 in Figure 5.2.) is the left most digit in the base 8 numeral.

In a sense, Algorithm I works from left to right, while Algorithm II works from right to left.

So, why bother with a new, harder-to-understand, algorithm, especially when we've gone to all the trouble of writing a program based on the other one? There are two answers, both a little subtle.

Let's address the "computational" answer first. So far, all of our experience has led us to place little value on a computer's time. Maybe our program takes 50 seconds to run. Is it worth spending an hour of our time to produce a better program that will run in 40 seconds? What if we were dealing with a large mainframe computer that cost $10 a

	Remainder
8 |2001	
250	1
8 |250	
31	2
8 |31	
3	7
8 |3	
0	3
	3721 (base 8)

Figure 5.2

second to run? (As long as my salary is less than $100 per hour, it makes sense to rewrite the program. If I have to pay for the computer time myself, it certainly makes sense!) What if, instead of seconds, we were talking about hours? There are many important problems from pure mathematical research to Social Security, from energy research to flight technology, that cannot be RUN on (even the fastest) computers because no one has yet discovered a "fast" enough algorithm. It is not too early to develop good habits. Look for the best algorithm you can find.

The other answer is mathematical. It is very often the case that one can achieve a better understanding of something by looking at it in different ways, from different points of view. Surely, you wouldn't buy a car without walking around it to see what it looks like from several directions. As the mathematics you study becomes more sophisticated, you will find this same approach of looking at things from different angles to be useful and rewarding.

What about doing some arithmetic, say in base 8. What is 5(base 8) + 7(base 8). Of course, 5 and 7 are the same in base 8 as they are in base 10. Moreover, five plus seven = twelve, in *any* number base. The addition of numbers is independent of what we choose to call the numbers. Five eggs + seven eggs is a dozen eggs in any language. What we are really asking then, is the numeral for twelve in base 8. It's 14 (a single eight plus four ones). Here is the addition table for base 8:

+	0	1	2	3	4	5	6	7
0	0	1	2	3	4	5	6	7
1	1	2	3	4	5	6	7	10
2	2	3	4	5	6	7	10	11
3	3	4	5	6	7	10	11	12
4	4	5	6	7	10	11	12	13
5	5	6	7	10	11	12	13	14
6	6	7	10	11	12	13	14	15
7	7	10	11	12	13	14	15	16

Sometime, generally between kindergarten and third grade, you stopped doing addition on your fingers because, by then, you had memorized the (ordinary) addition table for base 10 numerals. Meanwhile, the Vegan children memorize the above. The sum "5 + 4 = 11" is as natural for them as "5 + 4 = 9" is for you. Notice that their table is symmetrical just as ours is, reflecting the fact that addition of numbers (regardless of their names) is commutative.

Of course, in base 10, we have an algorithm for adding any two numbers, once we've memorized how to add two single-digit numbers. Does the algorithm still work in base 8? Take, for example, 45(base 8) + 67(base 8). (Follow along on scratch paper as you read the rest of this paragraph.) If we add 5 and 7, we get 14(base 8). What if we write down the 4 and "carry" the 1? Then 4(base 8) + 6(base 8) + the one we carried = 13(base 8). Is the answer 134(base 8)? How do we know?

We can, of course, convert the problem to base 10: 45(base 8) = 37(base 10) and 67(base 8) = 55(base 10). Now, in our old, familiar arithmetic, 37 + 55 = 92. Is 134(base 8) = 92 (base 10)? Well, we have a choice of whether to convert 134 (base 8) to base 10, or 92(base 10) to base 8. The easy way is the first: $1*64 + 3*8 + 4*1 = 92$, sure enough. While one example doesn't constitute a proof, it turns out that the familiar addition algorithm works as well for base 8 (or any other number base) as it does for base 10, provided we make the obvious modifications. In this case, "the modifications" amount to using the table above to add single digit numbers.

In conclusion, here is the addition table for base 2:

+	0	1
0	0	1
1	1	10

Exercises (5.2)

1. Complete the following base 8 multiplication table:

*	1	2	3	4	5	6	7
1	1	2	3	4	5	6	7
2	2						
3	3						25
4	4						
5	5						
6	6	14					
7	7						61

2. All the numbers in this problem are expressed in base 8. Do the indicated arithmetic (in base 8) and express the answer(s) in base 8.
   a. 16 + 61   b. 64 + 55   c. 73 + 25   d. 7 + 7 + 7
   e. 14 − 6   f. 76 − 67   g. 4*7   h. 10*10

3. All the numbers in this problem are expressed in base 2. Do the indicated arithmetic (in base 2) and express the answer in base 2.
   a. 10111 + 11011   b. 10011 + 11110   c. 11111 + 11111
   d. 11011 − 10111   e. 10000 − 1111   f. 10*10

4. When you memorized the (base 10) multiplication table, two rows were easy, row 5 and row 9.[†] Explain why. What rows of the base 8 multiplication table show similar properties? Explain.

5. Convert these base 10 numerals to base 8.
   a. 64   b. 73   c. 668   d. 888

6. Convert these base 10 numerals to base 2.
   a. 64   b. 73   c. 100   d. 1023

7. Convert these base 8 numerals to base 2.
   a. 10   b. 77   c. 22   d. 111

8. Using Algorithm II, write a program to convert from base 10 to base 8.

9. Write a program to convert from base 10 to base 2. (*Hint:* It's likely that a DIM statement will be needed.)

10. The usual algorithm for multiplying two several digit numbers together works in base 8 (provided you know how single digit numbers multiply ... see Exercise 1). Try to do these problems entirely in base 8. Show your work.
    a. 17*17   b. 75*57   c. 34*16   d. 37*42

11. Consider base B, where B is an integer, 1<B<10. Explain why
    a. 1(base B) = 1 (base 10).   b. 10(base B) = B(base 10).

---

†The 9-by-9 base 10 multiplication table contains 81 numbers. The 7-by-7 base eight multiplication table contains only 49. How many numbers does the base 2 multiplication table contain?

5.3 STRINGS

Let's begin with the notion of a "string variable." Apart from CHR$ and ASC, we have not addressed the concept of a "string." We have been using the computer exclusively for numerical computations, largely ignoring its "word processing" capabilities.

Recall that a variable name is associated with a memory location. Apart from subscripts,[†] a variable name can have one or two characters, the first of which must be a letter from A to Z. You can use a variable name having more than two characters, but only the first two are recognized by most microcomputers. For example, the typical micro won't distinguish AB and ABD, or R12 from R1. However, AB(12) is a different story.

A text "string" is just a sequence (or string) of symbols, like a word. Text strings may also be stored in memory locations and labeled with variable names. But, it is necessary to inform the computer in advance, what kind of "animal" is to be stored in its big zoo of a memory. To indicate that a memory location is to be reserved for a text string, the variable name must end with the symbol "$". (In an array, the "$" comes after the "array name", but before the subscripts.) That is, A$, AB$, A1$, A1$(I), C2$(5,3), and ABC$(17) are all legitimate string variable names.

Experiment a bit. RUN these.

```
10 A$="PETER LOVES PATTY"
20 PRINT A$
30 END

10 B$(1)="JOHN, "
20 B$(2)="JIM, "
30 B$(3)="AND FRANK"
40 FOR I=1 TO 3
50 PRINT B$(I);
60 NEXT I
70 END
```

[†] That is, array variables.

Note the use of quotes. What happens when you ENTER this

```
A$=COMPUTER
```

with no quotes (TYPE MISMATCH ERROR), or this

```
A(3)$="B"
```

(SYNTAX ERROR).

Other problems occur when you attempt to use BASIC code words as variable names. Try ENTERing

```
IF$="C3PO"
```

(SYNTAX ERROR), and

```
RUN$="R2D2"
```

(UNDEF'D STATEMENT ERROR).

RUN this program for a little surprise.

```
10  N$="123"
20  M$="456"
30  PRINT N$+M$
40  END
```

Character strings can be "added" in the sense of connecting them to form a combined string. (The fancy word for this kind of combination is "concatenation.") Another observation to be made about this example is that digits can be used in strings, just as if they were letters. Other symbols can be used, too. On the other hand, don't expect digits in strings to be treated as numbers. The computer recognizes the 5 in A$ = "C2:5B" as a letter, not a number. Sometimes this can be inconvenient. So, there is a BASIC command to convert a character string consisting of digits into a number. It is the VAL command. VAL([string]) returns the numeric VALue of the string. If the first (non-blank) character is not a plus (+), minus (−), or digit, the VALue returned will be zero. Here is a program to demonstrate VAL.

```
10  A$ = "−123"
20  B$ = "456"
30  PRINT A$+B$
40  PRINT VAL(A$)+VAL(B$)
50  END
```

The VALue of a mixed string like A\$ = "17A;B5" will be 17. String conversion ceases at the first non-digit character. Try it. Before RUNning the next program, try to figure out what the output will be.

```
10  FOR  I=1  TO  5
20  W$=W$+CHR$(64+I)
30  X$=X$+CHR$(48+I)
40  NEXT  I
50  PRINT  W$+X$
60  END
```

The opposite of VAL is STR\$. This command returns the STRing representation of the number in the argument. For example, STR\$(65) is the string, "65". One annoying aspect of the program developed in Section 5.2 to convert from base 10 to base 8 is that the output looks like a collection of digits, rather than a numeral. While this is not particularly important, it does provide an opportunity to illustrate STR\$ and VAL. If you replace the output lines, 130–160, with these, the problem will be corrected.

```
130  Q$=STR$(R(1))
140  FOR  I=1  TO  P
150  Q$=STR$(Q(I))+Q$
160  NEXT  I
170  PRINT  VAL(Q$)
```

Let's return to our discussion of number bases. We began by wondering what numerals might be used by Vegans, people with four fingers on each hand. What about the Procyons, with six fingers on each hand? Let's give base 12 a look.

The Procyons will, of course, have developed twelve individual digits, 0, 1, 2, ..., 8, 9, and symbols for ten and eleven. What would the symbols look like? We have no more idea about that than the Vegans have about "8" and "9"! We could, of course, use any symbols to represent ten and eleven; let's make them T and E.

Consider the base 12 numeral 3TE1. What does the number look like in base 10? Easy! We have a 1 in the one's column, an E in the twelve's column, a T in the $12^2 = 144$'s column, and a 3 in the $12^3 = 1728$'s column. Thus

$$3TE1(\text{base } 12) = 3*1728 + 10*144 + 11*12 + 1*1$$

$$= 5184 + 1440 + 132 + 1 \qquad (5.2)$$

$$= 6757(\text{base } 10)$$

Let's write a program to convert from base 12 to base 10:

```
10 INPUT "HOW MANY DIGITS ARE THERE";D
20 PRINT "ENTER THE BASE 12 NUMERAL ONE"
30 PRINT "DIGIT AT A TIME."
```

Now we're ready to start INPUTting the digits. We would like to be able to INPUT digits like 3, and "digits" like T. Unfortunately, that causes a difficulty in the program. Do we ENTER

```
60 INPUT A
```

or

```
60 INPUT A$
```

One way out of the difficulty is to just INPUT 10 instead of T. The computer won't know that 10 isn't a digit. Still, that solution seems to adulterate the basic idea. It is a "mixed" solution in which we do some of the computer's work. It is not a fully automated solution. Here is an alternative:

```
40  N=0
50  FOR I=D TO 1 STEP -1
60  INPUT A$
70  IF A$="T" THEN A=10
80  IF A$="E" THEN A=11
90  IF A$<>"T" AND A$<>"E" THEN A=VAL(A$)
100 N=N+A*12↑(I-1)
110 NEXT I
120 PRINT "THE BASE 10 NUMERAL IS"
130 PRINT N
140 END
```

Use this program to confirm each of the following:

$$10(\text{base } 12) = 12(\text{base } 10)$$

$$\text{TE}(\text{base } 12) = 131(\text{base } 10)$$

$$\text{TOTE}(\text{base } 12) = 17,411(\text{base } 10)$$

Exercises   (5.3)

1. Do these problems entirely in base 2.

a. 10111   b. 11001   c.   1011
   ×11   ×101   ×111

2. Confirm your arithmetic in Exercise 1 by means of the following algorithm:

   a. Convert the base 2 numerals to base 10.
   b. Do the multiplication in base 10.
   c. Convert the base 10 answer to base 2.
   d. Compare the result with the answer you got in Exercise 1.

3. Describe the difference in output of the following two programs:

```
10 A=12:B=34        10 A$="12":B$="34"
20 PRINT A;B        20 PRINT A$+B$
30 END              30 END
```

4. Describe what happens when you ENTER
   a. PRINT VAL(STR$(15))
   b. PRINT VAL(STR$("15"))
   c. PRINT STR$(VAL("15"))
   d. PRINT STR$(VAL(15))

5. Convert these base 12 numerals to base 10.
   a. 1T   b. T0E   c. T1E   d. T1TE

6. Convert these base 11 numerals to base 10.
   a. T0   b. T00   c. T0T   d. T00T

7. Write a program to convert base 11 numerals to base 10.

8. What output will this program produce?

```
10  A$=""
20  FOR I=1 TO 12
30  READ N
40  A$=A$+CHR$(N)
50  NEXT N
60  PRINT A$
70  DATA 71,79,79,68,32,77
80  DATA 79,82,78,73,78,71
90  END
```

9. Explain how the output from the program in Exercise 8 would change if line 40 were replaced with

```
40  A$=CHR$(N)+A$
```

10. Explain how the output from the program in Exercise 8 would change if line 40 were replaced with

    40 A$=A$+STR$(N)

11. Here is a number trick: Start with any 3-digit (base 10) number, call it XYZ. Form the (concatenated) number N = XYZXYZ. Divide N by 7, and call the quotient Q7. (Surprise #1: No matter what 3-digit number you started with, there is no remainder!) Divide Q7 by 11; call the remainder Q11. (#2: No remainder.) Finally, divide Q11 by 13. The quotient is your original number, XYZ!
    a. Write a computer program to
         (i) INPUT 3 digits, X, Y, and Z.
        (ii) Check to see that X $\neq$ 0.
       (iii) Concatenate to form N = XYZXYZ.
        (iv) Divide N by 7 and display the quotient.
         (v) Divide the quotient by 11 and display the new quotient.
        (vi) Divide by 13 and confirm the trick.
    b. Explain how the trick works.
    c. Can you devise a similar trick for 4-digit numbers?
    d. Why do 7, 11, and 13 add spice to the 3-digit trick?

12. The "binary," or base 2 system, is very important in computer science. Other numeration systems of interest to computer scientists (apart from base 10) are the "octal," base 8, and the "hexadecimal," base 16. Denote the single digit numerals, base 16, by 0, 1, . . . , 8, 9, A, B, C, D, E, and F. Then, for example, C(base 16) = 12(base 10). By hand, without using the computer, convert these base 16 numerals to base 10.
    a. 10   b. 100   c. AE   d. 1F

13. In base 16 (see Exercise 12), what is the next numeral after
    a. F   b. AF   c. 1E   d. FF

14. Write a computer program to convert from the hexadecimal (sometimes known briefly as HEX) system to the base 10 system. (*Hint:* See Exercise 12.)

15. Convert the following hexadecimal numerals to base 10. (*Hint:* See Exercise 14.)
    a. D3A8   b. A1C3EF   c. 8E4C93   d. 88C310

5.4 LOGARITHMS

Consider this table of powers of 2

Power	1	2	3	4	5	6
Value	2	4	8	16	32	64
Power	7	8	9	10	11	$\cdots$
Value	128	256	512	1024	2048	$\cdots$

We have become familiar with the ideas of exponents, essentially of reading this table from top to bottom. If we want to know what 2^5 is, we are "thinking" top to bottom (in the 5th column). If we want to convert 1101(base 2) to base 10, we are "thinking" top to bottom:[†]

$$1*[2^3] + 1*[2^2] + 0*[2^1] + 1 = 13.$$

Suppose we want to convert base 10 to base 2. Then, in a sense, we are using the table from bottom to top. For example, to convert 75(base 10) to base 2, we might proceed this way. Find the largest number in the Value row which is less than or equal to 75. It's 64. Now find the largest number in the second row $\leq 75 - 64 = 11$. It's 8. Since

$$11 - 8 = 3 = 2 + 1,$$

$$75 = 64 + 8 + 2 + 1$$

Now, using the table from bottom to top, $64 = 2^6$ and $8 = 2^3$. So,

$$75(\text{base } 10) = 1001011(\text{base } 2).$$

When using the table from bottom to top, it is usual to give the rows different names. Specifically, "Power" becomes "$\log_2$", the "logarithm" based on 2. Thus, "logarithm" is a fancy word for "power."[‡]

[†] Here it might make sense to reverse the direction of the table so that it becomes consistent with our writing of digits using place value.

[‡] Why use a fancy word when a simple one will do? Why complicate things unnecessarily? For the same reason we stubbornly cling to the complicated English system of measurement when the metric system is more sensible and easier to use. Just as a machinist with an expensive set of English tools won't welcome a change to the metric system, so the scientific community with its books, tables, and calculators "calibrated" for logarithms will resist changing, even to a more sensible word. *Our* position is that, in part, we are preparing to communicate in a scientific world. We have to be able to talk to people in the "outside" world. The hard fact is that, since they all use "logarithm," we're stuck with it, too.

Let's practice "taking" logarithms, remembering that it's just a matter of reading the table from bottom to top.

$$\log_2(32) = 5 \qquad \log_2(256) = 8 \qquad \log_2(2048) = 11$$

So, it's easy . . . *provided* we have a table handy. What about $\log_2(4096)$? Well, that's just the next power of 2 off the table, namely, 12, that is, $\log_2(4096) = 12$. Do you see the problem? Even if we do have a table before us, it may not be complete. But, more about that later.

Suppose you want to compute the product 16*32. Here is a new way to do it. Find 16 and 32 in the second row of the table. Look to the top row for their logarithms, 4 and 5. Now add $4 + 5 = 9$. Look along the top (Power) row until you come to 9. The number below 9 is the answer, namely, 512. Let's try 16*128. Their logarithms are 4 and 7; $4 + 7 = 11$, and $\log_2(2048) = 11$, so, $16*128 = 2048$. This simple idea of turning multiplication problems into addition problems is so cute that it is worth remembering. Indeed, it is nothing less than a new algorithm for multiplication. Here is a way to state the algorithm in brief:

$$\log_2(x) + \log_2(y) = \log_2(x*y) \qquad (5.3)$$

Notice how quickly a simple idea can become mysterious. Why is that? It's because we're lazy. We're too lazy to write out a 2-page description for our new algorithm. Instead, we write down a "shorthand" description. The trouble with any shorthand system is that, in order to make sense of it, one has to translate back to longhand. The shorthand is a code to be deciphered. We just have to be careful not to lose the key to the code. In this case, the "key" is the definition of "$\log_2$". Think of it this way:

$\log_2(x) = $ "The power of 2 that it takes to produce x."

Actually, there is another (mathematically equivalent) way to state Equation (5.3). It is

$$2^a * 2^b = 2^{a+b}. \qquad (5.4)$$

Our new algorithm for multiplication is a consequence of one of the ordinary rules of exponents.

In many ways, mathematics is like a game. Let's think in those terms for a minute. Here is a new game: Logarithms. We open up the box and look in the top for the rules. In mathematics, the role of "rules" is played by "definitions." In this case, the only "rule" is that $\log_2(x)$ is the power of 2 that it takes to make x.

The function of the rules is to tell us how to play the game, not how to play it *well*. Is it a defect in the rules of chess that upon reading them for the first time, one does not immediately become an expert? Of course not! And, no one expects that you are now an expert in logarithms, just because you've read the definition.[†] In mathematics, as with games, one becomes an expert by developing strategies. (Don't confuse strategies with rules. In mathematics, the term for "strategy" is "theorem.") One strategy for the logarithm game is Equation (5.3).

Here is another useful strategy, an extension of (5.3). (A theorem that follows easily from a previous one is called a "corollary.") Observe:

$$\log_2(x) = \log_2(\underbrace{x*x*\cdots*x}_{n\text{-times}})$$

$$= \overbrace{\log_2(x) + \log_2(x) + \cdots + \log_2(x)}^{n\text{-times}}$$

$$= n*\log_2(x),$$

by $(n-1)$ applications of strategy (5.3). Let's summarize this as

$$\log_2(x^n) = n*\log_2(x). \tag{5.5}$$

It's time we asked ourselves, "What's so special about 2 in all of this?" The answer is, simply, nothing. We could just as well have been talking about 3, or some other number. What, for example, is $\log_3(81)$? It is the power of 3 that it takes to make 81. Maybe that power is 2? No, $3^2 = 9$. Maybe it's 3? No, $3^3 = 27$. Maybe 4? Yes, $3^4 = 81$. So, $\log_3(81) = 4$, the power of 3 that it takes to make 81 is 4. It's too bad we don't have a table of powers of 3, or we could have seen at a glance what $\log_3(81)$ was, just by reading the table from bottom to top. Is it worth making up a $\log_3$ table? As sure as we do, someone will ask for $\log_4(256)$ or $\log_5(625)$.

[†] Even the word "definition" is used a little differently in mathematics. If you read the definition of a word in a dictionary, you expect to obtain a good feeling for what the word means. (If you don't, you might want to buy a better dictionary.) But, when you see a mathematical definition, it may not give you any kind of a feeling for what the word means. That's because mathematical definitions are to be compared with the rules of a game, not with dictionary definitions. You become familiar with mathematical definitions as you play the game.

On the other hand, no matter what "base," b, is chosen, $\log_b$ will exhibit properties of which Equations (5.3) and (5.5) are just examples. Namely,

$$\log_b(x) + \log_b(y) = \log_b(x*y) \qquad (5.6)$$

$$\log_b(x^n) = n*\log_b(x). \qquad (5.7)$$

(The word "base" as it's used with logarithms has nothing to do with "base" as it's used with numerals, that is, with number base.)

It turns out that the computer comes equipped with a library function called LOG. Try ENTERing

```
PRINT LOG(8)
PRINT LOG(81)
PRINT LOG(256)
PRINT LOG(625)
PRINT LOG(0)
PRINT LOG(-1)
```

These examples raise some interesting questions. For one thing, we can see that LOG is not $\log_2$, because if it were, LOG(8) would have been 3. Moreover, since LOG(81) isn't 4, LOG can't be $\log_3$. Similarly, LOG is neither $\log_4$ nor $\log_5$. (Explain why.)

LOG(0) caused a problem. Let's address it in the context of $\log_2$. Now, $\log_2(0)$ should be the power of 2 that it takes to produce 0; but, no power of 2 gives 0. Thus, there is no $\log_2(0)$. It doesn't exist. It is no wonder the computer had trouble; we asked it to produce a nonexistent entity. The same remarks hold for negative numbers. In strict, rather starchy, terms "the domain of definition" of any logarithm function is the set of positive numbers.

If LOG isn't $\log_2$, or $\log_3$, or . . . , what is it? For the moment, let's let e stand for the base of LOG. Then LOG $= \log_e$, where e is some mystery number. What good does LOG do us if we don't know what e is? The answer is that if we know one logarithm, we know them all. Here is how it works. Suppose you have $\log_a(x)$ and you want $\log_b(x)$. Using Equation (5.7), with different notation, we have

$$\log_a(b^n) = n*\log_a(b) \qquad (5.8)$$

Suppose we choose to let $n = \log_b(x)$. Then the right-hand side becomes

$$\log_b(x)*\log_a(b) = \log_a(b)*\log_b(x) \qquad (5.9)$$

(Logarithms are just numbers, and multiplication of numbers is commutative.) What about the left-hand side? Let's evaluate it from the inside out. How do we determine $b^{\log_b(x)}$? The only way we can. We are playing a game with just one rule! If we can't base our argument on that rule, we can't expect to get anywhere. The rule is that $\log_b(x)$ is the power of b that it takes to produce x. Suppose we raise b to the power of b that it takes to produce x. Should we be surprised to produce x? In other words,

$$b^{\log_b(x)} = x$$

It follows that the left-hand side of (5.8) becomes $\log_a(x)$. Comparing with (5.9), we get the nice formula

$$\log_a(x) = \log_a(b)*\log_b(x), \tag{5.10}$$

just as if the b's were to cancel. Alternatively,

$$\log_b(x) = [\log_a(x)]/[\log_a(b)] \tag{5.11}$$

All right. In a sense, we do know what $\log_a(x)$ is for one a, at least our computer does, namely, $a = e$. $\text{LOG}(X) = \log_e(X)$. We may not know much about e, but that doesn't matter in either (5.10) or (5.11). It's only the *values* of $\text{LOG}(X)$ that we need, not how they're obtained. Indeed,

$$\log_b(x) = \text{LOG}(x)/\text{LOG}(b) \tag{5.12}$$

Try these

```
PRINT LOG(32)/LOG(2)
PRINT LOG(9)/LOG(3)
PRINT LOG(256)/LOG(4)
PRINT LOG(125)/LOG(5)
```

Do you get $\log_2(32)$, $\log_3(9)$, $\log_4(256)$, and $\log_5(125)$?

Exercises (5.4)

1. By hand, without using the computer, evaluate:
   a. $\log_2(64)$ b. $\log_4(64)$ c. $\log_8(64)$ d. $\log_{64}(64)$

2. By hand, without using the computer, evaluate:
   a. $\log_6(36)$ b. $\log_8(512)$ c. $\log_{12}(1728)$ d. $\log_5(3125)$

3. By hand, without using the computer, evaluate:
   a. $\log_2(4096)$ b. $\log_4(4096)$ c. $\log_8(4096)$ d. $\log_{16}(4096)$

4. Given that $b^1 = b$, for any b, show that $\log_b(b) = 1$, for any b.

5. Suppose a and b are positive numbers bigger than 1. Prove that $\log_a(b) = 1/[\log_b(a)]$. (Hint: *Use* Equation (5.11) and Exercise 4.)

6. Use Equation (5.12) and the computer to confirm your answers to Exercise 1. (*Hint:* $\log_2(64) = \text{LOG}(64)/\text{LOG}(2)$.)

7. Use Equation (5.12) and the computer to confirm your answers to Exercise 2.

8. Use Equation (5.12) and the computer to confirm your answers to Exercise 3.

9. Evaluate:
   a. LOG(2)   b. LOG(3)   c. LOG(4)   d. LOG(6)
   e. LOG(9)

10. By hand, without using the computer, evaluate $\log_2(4)$, $\log_2(8)$, $\log_2(32)$, and $\log_2(64)$. Use your answers to confirm that:
    a. $\log_2(4*8) = \log_2(4) + \log_2(8)$.   b. $\log_2(4^3) = 3*\log_2(4)$

11. Using your answers to Exercise 9, show that:
    a. LOG(4) = LOG(2) + LOG(2).
    b. LOG(6) = LOG(2) + LOG(3).
    c. LOG(9) = 2 LOG (3).

12. Write a program to INPUT b, x and output $\log_b(x)$.

13. Solve for x if
    a. $2^x = 16$ b. $4^x = 16$ c. $2^x = 2$ d. $1^x = 2$

14. What do you think $\log_b(2)$ would be if b were 1?

15. Observe that LOG(1) = 0. Explain why $\log_b(1) = 0$ for every $b>0$.

16. Prove that $\log_b(x/y) = \log_b(x) - \log_b(y)$.

17. Prove that:
    a. $\log_b(1/2) = -\log_b(2)$ b. $\log_b(1/10) = -\log_b(10)$.

5.5 THE GEOMETRIC MEAN

Consider, once again, the table of powers of 2:

y	1	2	3	4	5	6	7	8
2^y	2	4	8	16	32	64	128	256

With this labeling of the rows, one expects the table to be read from top to bottom. To find 2^6, go along the first row, "row-y", until you find 6. Then drop down to the second row and find $64 = 2^6$. When it is expected that the table will be read from bottom to top, it is common to label the rows differently:

$\log_2(x)$	1	2	3	4	5	6	7	8
x	2	4	8	16	32	64	128	256

The most important thing to remember is that *the table is the same*, regardless of what labels are put on the rows.

We can have some fun with the notation. Notice

$$x = 2^y \text{ if and only if } y = \log_2(x). \qquad (5.13)$$

Or, to put it another way,

$$x = 2^{\log_2(x)} \text{ and } \log_2(2^y) = y \qquad (5.14)$$

We observed in the last section that logarithm tables are generally incomplete. To find, for example, $\log_2(4096)$, we need, in principle, to extend the table a few places to the right. But what about $\log_2(3)$? What power of 2 does it take to produce 3? How do we "extend the table" to repair this kind of deficiency?

As a matter of fact, why should there even be such a power of 2? How do we know that there is any solution, y, to the equation $2^y = 3$? Well, for one thing, we know that

$$\log_b(x) = \text{LOG}(x)/\text{LOG}(b). \qquad (5.15)$$

Therefore, not only does $\log_2(3)$ exist, but we have the means to find it.

$$\log_2(3) = \text{LOG}(3)/\text{LOG}(2)$$

$$= 1.58496\ldots$$

The power of 2 that it takes to produce 3 is $1.58496\ldots$. So, for example, we can fill in a bit more of our table:

y	1	1.58...	2	3	4	5	6
2^y	2	3	4	8	16	32	64

Similarly, we can find $\log_2(5)$, $\log_2(6)$, $\log_2(7)$, ..., and so on. We can fill in as much of the table as we want. But, wait a minute. What, in plain English, does $2^{1.58496}$ mean? Let's work out a simpler example first, say $2^{.5}$, or $2^{1/2}$. Recall the fundamental rule for exponents:

$$2^x * 2^y = 2^{x+y} \tag{5.16}$$

If we let $x = y = 1/2$ in Equation (5.16), we see that $2^{1/2}$ times itself is 2. In other words, $2^{1/2}$ is the square root of 2. Let's use the computer to confirm this observation. ENTER each of these

```
PRINT 2↑(1/2)
PRINT 2↑.5
PRINT SQR(2)
```

It must also be that $4^{1/2}$ is the square root of 4, namely, 2. Confirm this by ENTERing

```
PRINT 4↑(1/2)
PRINT 4↑.5
```

What about $2^{1.5}$? With $x = 1$ and $y = .5$ in Equation (5.16), we obtain

$$2^1 * 2^{.5} = 2^{1.5}.$$

That is, $2^{3/2} = 2 * 2^{1/2}$. It is, perhaps, more revealing to write it as $2^{(1/2)3}$, that is, $2^{3/2}$ is $(\sqrt{2})^3$. It is also $\sqrt{2^3}$.

Now, we're ready for $2^{.3} = 2^{3/10}$. We can think of it either as the third power of $2^{1/10}$, or the 10th root of 2^3. That is, $2^{2/3}$ is [the 10th root of 2]^3, or the 10th root of 8. Let's use the computer to confirm it:

```
PRINT 2↑(3/10)
PRINT 2↑.3
PRINT (2↑(1/10))↑3
PRINT (2↑3)↑(1/10)
```

Once again, 2 has been used in an illustrative way. Nothing depends on it. Let's do one final example: Compute $5^{3.14}$. First of all, write this as $5^3 * 5^{.14}$. Since $5^3 = 125$, it remains to evaluate $5^{.14}$; but, $.14 = 14/100$. Thus, $5^{.14} =$ the 100th root of 5^{14}, or the fourteenth power of the 100th root of 5. To compute an actual value, we may ENTER any one of these commands:

```
PRINT  5↑3.14
PRINT  125*5↑.14
PRINT  125*(5↑(1/100))↑14
PRINT  125*(5↑14)↑.01
```

We can apply some of these ideas to produce a new average called the *geometric mean*. The (ordinary) mean[†] is computed by means of addition; but, addition isn't the only operation available. What about trying to base an average on multiplication?

Suppose you are given a data set with two elements, a and b. Their mean is $(a + b)/2$. Think of the mean as a number m such that $(m + m)/2 = (a + b)/2$, that is,

$$m + m = a + b$$

If there were three numbers, the comparable equation would be

$$m + m + m = a + b + c$$

Presumably, a *multiplicative average* would have the corresponding property, namely,

$$m*m = a*b,$$

or

$$m*m*m = a*b*c.$$

In the first case, m is the square root of $a*b$. In the second, m is the cube root of $a*b*c$. In general, the geometric mean of n positive numbers is the n-th root of their product.

There is a famous inequality concerning these two means. The *Arithmetic-Geometric Mean Inequality* asserts that the (arithmetic) mean of n positive numbers is strictly greater than their geometric mean unless

---

[†]The ordinary mean is sometimes called the "arithmetic" mean to distinguish it from the geometric mean.

the n numbers are all the same. Here is a proof of the inequality in case $n = 2$: Suppose $a \neq b$. Then

$$0 < (\sqrt{a} - \sqrt{b})^2$$

$$0 < a - 2\sqrt{ab} + b$$

$$2\sqrt{ab} < a + b$$

In conclusion, let's return for a moment to exponents. We have seen how to deal with fractional exponents. In Section 1.3, we discussed negative exponents. Recall, for example, that $2^{-1} = 1/2$, $2^{-2} = 1/2^2$ $= 1/4$, $2^{-3} = 1/2^3 = 1/8$, etc. It follows that

$$\log_2(1/2) = \log_2(.5) = -1,$$

$$\log_2(1/4) = \log_2(.25) = -2,$$

$$\log_2(1/8) = \log_2(.125) = -3,$$

and so on. More generally,

$$\log_b(1/x) = \log_b(x^{-1}) = -\log_b(x)$$

For $b > 1$, the logarithm of a positive number less than 1 will be negative. That is,

$$\log_b(x) \text{ is } \begin{cases} > 0, & \text{if } x > 1 \\ = 0, & \text{if } x = 1 \\ < 0, & \text{if } 0 < x < 1 \end{cases} \tag{5.17}$$

Remember that $\log_b(x)$ is undefined if $x \leq 0$.

Exercises (5.5)

1. Compute $\log_2(5)$, $\log_2(6)$, $\log_2(7)$, $\log_2(9)$, and $\log_2(10)$. Use these values to:
   a. fill the gaps in the $\log_2(x)$ table from $x = 1$ to $x = 10$.
   b. confirm that $\log_2(10) = \log_2(2) + \log_2(5)$.
   c. confirm that $\log_2(9) = 2*\log_2(3)$.

2. Fill in the missing values in the following table:

$\log_3(x)$	0	___	1	___	___
x	1	2	3	4	5

$\log_3(x)$	___	___	___	2	___
x	6	7	8	9	10

3. Use your $\log_3$ table (from Exercise 2) to confirm that:

a. $\log_3(10) = \log_3(2) + \log_3(5)$
b. $\log_2(3) = 1/\log_3(2)$

4. By hand, without using the computer, calculate
a. $25^{1/2}$ b. $8^{1/3}$ c. $32^{1/5}$ d. $64^{1/6}$ e. $64^{1/3}$

5. By hand, without using the computer, calculate
a. 5^{-1} b. 10^{-2} c. $(1/2)^{-1}$ d. $.5^{-1}$ e. $.5^{-3}$

6. By hand, without using the computer, calculate:
a. $8^{2/3}$ b. $27^{1/3}$ c. $27^{2/3}$ d. $32^{2/5}$ e. $32^{3/5}$

7. By hand, without using the computer, calculate
a. $8^{-1/3}$ b. $8^{-2/3}$ c. $16^{-3/4}$ d. $32^{-4/5}$ e. $(1/4)^{-5/2}$

8. Compute:
a. $\log_2(1/4)$ b. $\log_3(1/4)$ c. $\log_4(1/4)$ d. $\log_3(4/5)$

9. Compute:
a. $\log_4(2)$ b. $\log_8(2)$ c. $\log_8(4)$ d. $\log_{81}(27)$

10. Given a positive number N, write a program to find the smallest whole number K such that $K>\log_2(N)$. (*Hint:* This is the same problem as Exercise 3, Section 2.4. See if you can improve on the solution to that exercise.)

11. Consider changing the dimensions of the search area in the "SUTHERLAND" game of Section 4.1. Suppose the volume of space concealing the enemy craft ranges from GSC (0,0,0) to (N,N,N), where $N>15$. In order to keep the game "fair," you should also increase the number of barrages permitted in line 40. Explain what bearing Exercise 10 has on the situation.

12. Write a program to INPUT a positive number N and output its R-th roots, for R = 1, 2, . . . , 10, and 100. RUN your program for the following values of N:
a. N = 1   b. N = 2   c. N = 10   d. N = 0.5   e. N = 0.1

13. The 100th root of each number N in Exercise 12 is very close to which integer?

14. Compute the arithmetic mean, the geometric mean, and the RMS average of the numbers:
    a. 1, 1, 1, 16   b. 2, 2, 2, 2   c. 2, 4, 8, 16, 32

15. Write a program to INPUT or READ a set of positive numbers and output their geometric mean.

5.6 THE MYSTERY NUMBER

It is time we returned to the "mystery number," e, of Section 5.4. It turns out that

$$e = 2.718281828459\ldots$$

It may seem strange to base the computer's LOG function on such a number. Why go out of the way to make things difficult? Why not make a more natural choice, pick a simple whole number for b and program the resulting $\log_b$ into the computer? Ironically, that solution would cause *more* work for you, not less. In fact, e turns out to be the simplest, most natural base for a logarithm. Unfortunately, there is no way to justify these statements until you've had a chance to learn some integral calculus.

The fact that e seems like a funny number to be the base of the "natural logarithm" doesn't change the fact that $\log_e$ *is a logarithm.* In particular, $\log_e(x)$ is the power of e that it takes to produce x,

$$x = e^y \text{ if and only if } y = \log_e(x), \tag{5.18}$$

$$x = e^{\log_e(x)}, \text{ and } \log_e(e^y) = y \tag{5.19}$$

Because it is so important, $\log_e$ is usually distinguished from other logarithms by a special notation. Define $\ln(x) = \log_e(x)$ for all positive values of x.[†] Then, LOG(x) returns the value of $\ln(x)$.

[†]It is incorrect to use "$\ln_b$" or "$\ln_e$." The notation "ln" is a substitute for "$\log_e$."

Another library function is called EXP. It is defined by EXP(x) = e^x. In particular, from Equations (5.18) and (5.19), we see that LOG and EXP are inverses of each other; for example,

$$\text{EXP(LOG}(x)) = x, \; x>0, \tag{5.20}$$

and

$$\text{LOG(EXP}(y)) = y, \text{ for all } y. \tag{5.21}$$

Confirm these by ENTERing

```
PRINT EXP(LOG(1.2345))
PRINT LOG(EXP(8.9012))
```

Since EXP(1) = e^1 = e, you can confirm the first few digits of the value of e by PRINTing EXP(1).

In Section 1.4, we discussed compound interest. Recall that the return on P dollars invested for t years, at an interest rate r (expressed as a decimal), is:

$$P*[1+(r/m)]^{mt},$$

where m is the number of compounding periods in a year. We found, for example, that the return on $1000 invested for a (365-day) year at 12%, compounded daily, is $1127.47.

What if we were to compound the interest "continuously?" That is, suppose the interest is always being compounded? Let m increase to infinity in the expression

$$[1+(r/m)]^{(mt)}$$

The result, believe it or not, is e^{rt}. If P dollars is invested at the rate r, compounded continuously, then the value of the investment grows "exponentially" in t. After t years, it is $P*\text{EXP}(r*t)$. Let's use this formula to compute the return if P = 1000, r = .12, and t = 1. Then:

$$Pe^{rt} = P*\text{EXP}(r*t)$$

$$= 1000*\text{EXP}(.12)$$

$$= 1127.50$$

Over the course of a year, continuously compounded (12%) interest yields 3 cents more per $1000 than interest compounded daily. (Is it worth driving to the next county to find a bank that compounds continuously?)

How long would it take for your principal to double in an account in which interest is compounded continuously? After t years, the amount of money in your account would be

$$P(t) = P(0)e^{rt} \tag{5.22}$$
$$= P(0)*\text{EXP}(r*t),$$

where $P(0)$ is the value at $t = 0$, the initial deposit. We are after the value of t for which $P(t) = 2*P(0)$. So, set

$$2*P(0) = P(0)*\text{EXP}(r*t)$$

and solve for t. After cancelling $P(0)$, we obtain

$$2 = \text{EXP}(r*t)$$

Making use of Equation (5.21), we can transform this into the equivalent equation,

$$\text{LOG}(2) = r*t,$$

that is,

$$t = \text{LOG}(2)/r$$

Since $\text{LOG}(2) = .6931\ldots$, we obtain the "Rule of 70: At an interest rate r, compounded continuously (or even monthly), the time it takes an investment to double is between $.69/r$ and $.70/r$ years." If you earn 10%, compounded frequently, your money will double every seven years. [†]

We can apply the same idea to inflation. Think of it as an interest applied to prices. At an annual inflation rate of 10%, prices double every seven years. During an average lifetime (70 years), constant 10% inflation would lead to 1024-fold price increases ($2^{10} = 1024$). What could have been bought for a penny would cost $10.24 seventy years later.

Many other natural phenomena follow the basic exponential law expressed in Equation (5.22). One of these is radioactive decay. Suppose we start with $P(0)$ pounds of a radioactive substance having a "half-life" t_h. [‡] The amount present at time t is

$$P(t) = P(0)*e^{kt}, \tag{5.23}$$

[†] The Rule of 70 is usually stated for r expressed as a percent. If r is left as a percent, the time to double is between $69/r$ and $70/r$.

[‡] The "half-life" is the time it takes for one-half of the specimen to "decay."

where k is a constant depending on t_h. To determine k, we use the definition of half-life: $P(t_h) = P(0)/2$. Hence,

or
$$P(0)/2 = P(0)*EXP(k*t_h),$$

$$1/2 = EXP(k*t_h)$$

Applying LOG to both sides, we obtain

$$\ln(1/2) = kt_h,$$

that is,
$$k = [\ln(1/2)]/t_h$$

$$= -\ln(2)/t_h,$$

because $\ln(1/2) = \ln(2^{-1}) = -\ln(2)$. Putting this value for k into Equation (5.23) we obtain the formula for radioactive decay:

$$P(t) = P(0)*EXP[(-.693/t_h)*t] \qquad (5.24)$$

One isotope of carbon (carbon-14) has a half-life of 5700 years. It is frequently used to date archeological samples. Suppose a chemist determines that 90% of the carbon-14 from an ancient campfire has decayed. How old is the site? To solve this problem, we appeal to Equation (5.24). Since 90% of the carbon-14 has decayed, 10% of it remains. So, the t of interest to us satisfies the equation $P(t) = .10*P(0)$. On the other hand, $P(t) = P(0)*EXP[(-.693/5700)*t]$. Setting these two expressions equal and cancelling $P(0)$, we obtain

$$.10 = EXP[-.000122*t].$$

Taking the natural logarithm of both sides,

$$\ln(1/10) = -.000122*t$$

But, $\ln(1/10) = -\ln(10) = -2.30. \ldots$ So,

$$t = 2.30/.000122$$

$$= 18,900 \text{ years.}$$

Exercises   (5.6)

1. Suppose you were to buy a $1000, one-year certificate of deposit at a bank. How much would you receive at the end of the year if the interest rate is:

   a. 11.5%, compounded continuously?
   b. 11.6%, compounded monthly?
   c. 11.75%, compounded quarterly?
   d. 12% simple interest (not compounded at all)?

2. Recall that the annual "yield" is the amount of simple interest that would generate the same total interest payment at year's end. Compute the yield for each part of Exercise 1. Which is the better investment?

3. Suppose 10 years ago, one of your great-aunts secretly invested $1000 for you at an interest rate r, compounded continuously. What is the current value of the investment if r =
   a. 5%   b. 5½%   c. 6%   d. 6½%   e. 7%
   f. 10%   g. 10½%   h. 11%   i. 11½%   j. 12%

4. In 1983 a steel cylinder, discarded by a Mexican hospital, became mixed with other scrap metal to be recycled. The resulting metal was sold to several companies, typically to become the chief structural component of various manufactured items. The original use of the cylinder was to shield radioactive cobalt-60. Unfortunately, the Co-60 was not removed before the cylinder was discarded. Thus, the widely distributed tables, chairs, and other items were contaminated. The half-life of Co-60 is 5¼ years.
   a. What fraction of the Co-60 will still be present in the recycled steel in the year 2003?
   b. How long will it take (starting in 1983) for 90% of the contaminating cobalt-60 to decay? (*Hint:* This problem is the reverse of part a. There, you knew the time and wanted the amount present. Here you know the amount present and want to know the time. Take LOGs.)
   c. How long will it take for 99% of the radioactive cobalt to decay? (Can you think of two ways to do this problem? If so, work out both solutions and confirm that they are the same.)

5. For $-1 < x < 2$, EXP(x) = e^x is approximated by the polynomial function $p(x) = 1 + x + x^2/2 + x^3/6$. Compare EXP($x$) and $p(x)$ for x =
   a. -0.5 b. 0   c. 0.5   d. 1   e. 1.5

6. The mystery number, e, occurs frequently in unexpected places. For example, the function whose graph is the "normal" curve from statistics is $[\text{EXP}(-x^2/2)]/\sqrt{2\pi}$. One way to approximate e is by the series,

$$S(k) = 1 + 1/(1!) + 1/(2!) + 1/(3!) + \cdots + 1/(k!),$$

that is, e is approximately equal to $S(k)$, and the approximation gets better and better the larger k becomes. (As 0! is defined to be 1, the first term could be written as 1/(0!).) Write a program to input k and output $S(k)$. Use your program to compute $S(k)$ for
   a. $k = 2$. (Do this one by hand, and compare your answer with the computer's.)
   b. $k = 5$
   c. $k = 6$
   d. $k = 7$
   e. $k = 8$
   f. $k = 9$
   g. $k = 10$

7. In calculus, it is shown that the sum

$$s(k) = 1/1 + 1/2 + 1/3 + \cdots + 1/k$$

is approximately $\ln(k)$. Write a program to INPUT k and output $s(k)$. Use your program to compute $s(k)/\ln(k)$ for $k = $
   a. 10   b. 100   c. 1000   d. 10,000

8. The element strontium has atomic number 38. This means there are 38 protons in the nucleus of a strontium (Sr) atom, and 38 electrons surrounding it. In addition to protons, the nucleus of a strontium atom contains anywhere from 43 to 59 neutrons. This gives various atoms of strontium atomic weights between 81 and 97. (The atomic weight of an atom is the number of protons plus the number of neutrons in its nucleus.) Atoms with the same atomic number but of different atomic weights are called "isotopes." Chemically, two isotopes of the same element are indistinguishable because chemical properties are determined only by the electron field of the atom, hence, only by the atomic number. The most common variety of strontium in nature is the stable (nonradioactive) isotope Sr-88.

Strontium is needed to make strong bones and healthy bodies. Cows, in particular, concentrate strontium in their milk for the benefit of their offspring. We, of course, intercept a great deal of that milk; and some of the Sr that it contains finds its way into our bones. One of the byproducts of atomic bombs is a radioactive isotope of strontium, Sr-90, with a half-life of 25 years. A treaty banning the testing of nuclear weapons in the atmosphere was signed in the early 1960s, primarily to stop the fallout of Sr-90. What fraction of the Sr-90 produced in 1960 (was or) will still be around in:

a. 1985?   b. 1990?   c. 2000?   d. 2100?

9. Strontium-85 has a half-life of 65 days. How long would it take for 99.9% of a sample of Sr-85 to decay?

10. Strontium-94 has a half-life of 2 minutes. What fraction of a sample of Sr-94 remains after a day?

11. The largest atomic number that occurs in nature is 92, corresponding to uranium. One isotope of uranium, U-235, has a half-life of 0.71 billion years. What fraction of an initial quantity of U-235 remains after

a. 4 billion years?   b. 5 billion years?   c. 6 billion years?
d. 7 billion years?   e. 8 billion years?

(*Hint:* Let $Q(t)$ be the quantity of U-235 present at time t, where t is measured in units of 1 billion years. Then $Q(t) = Q(0)*\text{EXP}(-t*\text{LOG}(2)/.71)$. We want the ratio, $Q(t)/Q(0)$.)

12. A second isotope of uranium (See Exercises 8 and 11.), U-238, has a half-life of 4.5 billion years. What fraction of an initial quantity of U-238 is still around after

a. 4 billion years?   b. 5 billion years?   c. 6 billion years
d. 7 billion years?   e. 8 billion years?

13. In one theory of stellar evolution, supernovas play a prominent role. Two features of this theory interest us here:

a. Elements with atomic number greater than 26 (iron) are created only during supernova explosions.

b. The shock waves from supernovas cause interstellar dust and gas to "condense" into star systems.

Suppose this theory were correct. Assume that our solar system was formed as a direct result of a supernova explosion. Suppose, further, that all the uranium present in the solar system was "deposited" by the same supernova that produced the system. Finally, assume that U-235 and U-238 were created in equal amounts by the supernova. Then, we can deduce the age of the solar system just by determining the ratio of U-235 to U-238 in uranium ore. That ratio turns out to be about 0.7%.

Use the results of Exercises 11 and 12 to estimate the age of the solar system. (*Hint:* Since we assume the initial quantities of U-235 and U-238 were the same, the ratio of U-235 to U-238 after 4 billion years would be [the answer to 11a] divided by [the answer to 12a]. Similarly, if the solar system is 5 billion years old, the observed ratio of U-235 to U-238 should be [the answer to 11b] divided by [the answer to 12b]. Compute the ratios of the answers for corresponding parts of Exercises 11 and 12, and determine which comes closest to 0.7%.)

14. James Stirling, an 18th Century Scottish mathematician, showed that

$$n! \doteq \sqrt{2\pi n}\ (n/e)^n$$

for large values of n. (Here "$\doteq$" means "approximately equal.") Compare 20! with the approximation given by Stirling's Formula (with $n = 20$).

15. Evidently, $(1 + 1/n)^n$ gets closer and closer to e as n gets larger and larger. Why? Evaluate $(1 + 1/n)^n$ for $n =$
    a. 10   b. 100   c. 1000   d. 10,000
    (*Hint:* See Section 1.4, Exercise 1.)

16. Phone several local banks, savings-and-loans, or thrifts. Find out what the current interest rate, compounding periods, and minimum deposits are for:
    a. passbook savings accounts
    b. interest checking accounts
    c. money market accounts
    d. a $10,000 certificate of deposit

Chapter 6
NUMBER THEORY

6.1 PRIME NUMBERS

A positive integer greater than 1 is *prime* if it cannot be factored as a product of two smaller integers.[†] An integer greater than 1 that is not prime is said to be *composite*. Thus, a composite integer, n, can be expressed as a product, $n = k*q$, where both k and q are positive integers less than n. The primes are 2, 3, 5, 7, 11, ...

What about a number like 2047? Is it prime or is it composite? Here we are with another new "game." The only rule is the definition in the first paragraph. What we would like are some strategies to simplify the "playing" of the game. Let's see if we can develop some as we go along.

Basically, we're trying to solve an equation with two variables, namely, $2047 = k*q$, with an additional condition. With no extra restrictions, there are lots of solutions. One is $k = 2$ and $q = 1023.5$. The side condition, of course, is that we want both k and q to be integers (less than 2047). Let's say that k *divides* n, and write $k \mid n$, if there is an integer q such that $n = k*q$. In other words, "divides" is short for "exactly divides," with no remainder. In particular, 2 does not divide 2047. Naturally, every integer, n, is "divisible" by 1 and n. We'll refer to these as trivial divisors. A nontrivial divisor is a *proper* divisor.

[†] It is only for convenience that we have chosen to exclude negative "primes" from our discussion.

Thus, 2 and 3 are the proper divisors of 6, while 1 and 6 are the trivial divisors. A second definition of prime can now be given. An integer $n>1$ is prime if it has no proper divisors.

But, is this getting us any closer to finding out if 2047 is prime? Let's get on with that task. One (not very imaginative) idea is simply to try all possible proper divisors. We have already seen that 2 does not divide 2047. (Indeed, 2047 is odd!) What about 3? Well, $3 \mid n$ if and only if 3 divides the sum of the (base 10) digits comprising n. Since 2 + 0 + 4 + 7 = 13 is not divisible by 3, neither is 2047. What about 4? It is a waste of time even to ask the question. If $4 \mid n$, then $n = 4*q = 2*(2q)$, i.e., $2 \mid n$. And, we already know that 2 does not divide n.

The argument involving 4 can be generalized. There is no point in checking to see if any composite number divides 2047. If $p \mid m$ and $m \mid n$, then $p \mid n$. We have obtained our first strategy for prime numbers:

If n does not have a proper *prime* divisor,   (6.1)

then n is prime.

Another way to say the same thing is this:

If n is composite,   (6.1′)

it has a proper prime divisor.

Returning to 2047, what about 5? No, 5 doesn't divide it because the last digit, 7, is neither 0 nor 5. After 5, the next prime is 7. Dividing 2047 by 7, we obtain a quotient of 292 and a remainder of 3. Thus, 7 does not divide 2047. Skipping the composite number 8, 9, and 10, we come to 11.

I confess to being tired of this game already! If we go on much longer, we'll run into the problem of not knowing whether the next prospective divisor is, itself, prime or not. It's time to stop grinding and do a little more thinking.[†] Suppose p were the *smallest* prime to divide 2047. How big could p be? If $2047 = p*q$, and p is the smallest prime divisor, then $p \leqslant q$. Moreover, $p*p \leqslant p*q = 2047$. That is, $p^2 \leqslant 2047$, or $p \leqslant$

[†]This process of alternately grinding and thinking is very common in mathematics. There is nothing like the prospect of a lot of grinding to stimulate some creative thinking.

$\sqrt{2047}$. This leads to an improvement on the strategy reflected in Definitions (6.1) and (6.1′):

If n is composite,

then it has a prime divisor $p \leq \sqrt{n}$.   (6.2)

Since $\sqrt{2047} = 45.24 \ldots$, we won't have to continue the above process for as long as we might have thought. The remaining primes less than 45 are 11, 13, 17, 19, 23, 29, 31, 37, 41, and 43. While this is still a formidable list, it is not impossibly long, especially with some help from the computer. Try RUNning this:

```
10  FOR  I=1  TO  10
20  READ  A
30  PRINT  A ; 2047 / A
40  IF  2047 / A = INT ( 2047 / A )  THEN  STOP
50  NEXT  I
60  DATA  11 , 13 , 17 , 19 , 23 , 29 , 31 , 37 , 41 , 43
70  END
```

The STOP command interrupts program execution. The idea is that we can "stop" if one of the primes in the DATA statement turns out to be a divisor of 2047. In that case, 2047 is not prime, and there is no point in continuing. (The computer indicates that it has STOPped by outputting something like "break in 40.")

If, for some reason, we wanted to continue program execution, we could ENTER the BASIC instruction CONT. This causes the computer to CONTinue from where it left off when it STOPped. (If program execution was interrupted in line 40 when you ran the above program, experiment with CONT.)

Why would you want to STOP and not END? What if 2047/A is not an integer, but the first nonzero digit in its decimal expansion occurs beyond the number of digits carried by the computer? Then, the machine will *think* that 2047/A is an integer, and it will STOP. We might discover the mistake by multiplying together the PRINTed values of A and 2047/A. We could then pass by the spurious "divisor" and CONTinue to search for real divisors.

Here is another potential problem: What if 2047/A really is an integer but, due to round-off error, the computer doesn't recognize it as one. (Maybe the computer comes up with 509.00000001 when 509 is the correct quotient.) Then program execution will not STOP. Hopefully, our suspicion will be aroused by the PRINTed numbers.[†]

STOP is even more versatile than we have yet seen. It is a way to return "control" to you. It takes the computer out of the "programming mode" and puts it into the "direct mode," but in such a way that program execution can be resumed. CONT is just one way to resume execution. If you do not wish to CONTinue from the STOPping point, you can ENTER an instruction to GO TO any line number you choose. You can even change the content of a memory location before resuming program execution. RUN this program, and ENTER CONT when it STOPs.

```
 10  FOR  I=1  TO  20
 20  PRINT  I ;
 30  IF  I=10  THEN  STOP
 40  NEXT  I
 50  GO  TO  100
 60  PRINT  "I  SEE"
100  END
```

Then RUN it again and ENTER this sequence of instructions when execution STOPs.

```
I=15
CONT
```

Finally, RUN the program a third time, and resume execution with this:

```
GO  TO  60
```

The GO TO statement can also be used to *begin* program execution. In the last experiment, execution was terminated (not interrupted) by the END statement following line 60. The computer has forgotten that it has been RUNning this program. It's almost as if the program had just been ENTERed. Instead of RUN, ENTER these instructions:

```
I=10
GO  TO  20
```

[†]The computer may be doing its best and still mislead us! Never forget that a computer is not infallible. Mistakes can occur even when a program is logically correct and properly ENTERed.

Recall that the RUN command automatically clears certain memory locations. Thus, I = 10 followed by RUN would cause memory location I to revert to the default value of 0 before program execution begins.

The CONT command can also be used following the interrupt procedure. Experiment with something like this:

```
10 FOR I=65 TO 90
20 PRINT CHR$(I)
30 NEXT I
40 END
```

As soon as program execution has begun, use the interrupt procedure. Then ENTER the CONT instruction.

There is one danger with STOP—CONT. If you change any part of the program while execution is suspended—for example, by adding a line, deleting a line, or even adding a semicolon at the end of a line— program execution cannot be CONTinued. You will see a CAN'T CONTINUE ERROR.

Exercises   (6.1)

1. How many even prime numbers are there?

2. Make a list of all the prime numbers less than 100.

3. Write a program to INPUT a number N<10,000, and determine whether it is prime or not. Assume the INT function is infallible. (*Hint:* $\sqrt{10,000} = 100$.)

4. Write a program to INPUT a number N<10,000 and determine whether it is prime or not. (Assume the INT function is not infallible. Prepare a list of possible divisors of N by PRINTing those primes P for which .0001>ABS(N/P − INT(N/P)).

5. Determine which of the following are primes. For each composite number, provide the smallest prime divisor.
   a. 1987   b. 8633   c. 8641   d. 3309   e. 4327
   f. 9973   g. 8357   h. 439   i. 9073   j. 9409

6. Write a program to PRINT every prime number between 1 and 1000. (Assume that INT is infallible.)

7. RUN your program from Exercise 6, and record the output.

8. For each positive integer n, define $B(n)$ to be the number of primes less than or equal to n. Write down $B(n)$ for $n = 1, 2, \ldots, 20$.

9. Write a program to PRINT $B(n)$ for $n = 1, 2, \ldots, 1000$. (See Exercise 8. *Hint:* See Exercise 6.)

   The "Prime Number Theorem" asserts that $n/\ln(n)$ is a good approximation for $B(n)$ for large n. (See Exercise 8.)

10. Compute the ratio of $B(n)$ to $n/\ln(n)$ when
    a. $n = 10$
    b. $n = 100$ (*Hint:* Exercise 2)
    c. $n = 1000$ (*Hint:* Exercise 9)

11. In 1893, a man by the name of Bertelsen computed $B(1 \text{ billion}) = 50{,}847{,}478$. Compute the ratio of $B(n)$ to $n/\ln(n)$ when $n = 1$ billion.

12. When n is "small," it has been suggested that $B(n)$ is better approximated by $P(n) = n/[\ln(n) - 1]$. Find the ratio of $B(n)$ to $P(n)$ when
    a. $n = 10$ b. $n = 100$ c. $n = 1000$ d. $n = 1$ billion

13. Explain why it doesn't make much difference whether we use $n/\ln(n)$ or $n/[\ln(n) - 1]$ as an approximation to $B(n)$ for very large n.

14. If you pick a number between 1 and n at random, what is the probability that it will be a prime number? Compute the approximate probability for
    a. $n = 10{,}000$ b. $n = 100{,}000$ c. $n = 1$ million

6.2 DIGITAL ROOTS

Let's see how to prove that a (base 10) number N is divisible by 3 if and only if the sum of its digits, call it S, is divisible by 3. A typical base 10 numeral can be "decomposed" as

$$N = A*1 + B*10 + C*100 + D*1000 + \cdots$$

Then

$$S = A + B + C + D + \cdots$$

Now, expressing each power of 10 in terms of its quotient and remainder upon division by 3, we can also write N this way:

$$N = A + B*(3*3+1) + C*(3*33+1) + D*(3*333+1) \cdots$$

$$= A + 3*(B*3) + B + 3*(C*33) + C + 3*(D*333) + D + \cdots$$

$$= (A + B + C + D + \cdots) + 3*(B*3 + C*33 + D*333 + \cdots)$$

Dividing S by 3, we get a quotient Q and a remainder R. Thus, $A + B + C + D + \cdots = 3*Q + R$. Continuing from above,

$$N = (3*Q + R) + 3*(B*3 + C*33 + D*333 + \cdots)$$

$$= 3*[Q + (B*3 + C*33 + D*333 + \cdots)] + R$$

So, R is the remainder when we divide N by 3. In other words, whether we divide S by 3 or N by 3 we are left with exactly the same remainder. But, $3 \mid S$ if and only if $R = 0$, and $3 \mid N$ if and only if (the same) $R = 0$. Q.E.D.

Let's take the argument one step further. We have just seen that $3 \mid N$ if and only if $3 \mid S$. But, by the same token, $3 \mid S$ if and only if 3 divides the sum of *its* digits! Suppose we sum the digits of N to obtain $S = S1$. Now sum the digits of $S1$ to get $S2$. Sum the digits of $S2$ to get $S3$, etc. Stop when you reach a number, Sk, consisting of only one digit. Then Sk is the *digital root* of N. By repeated application of the argument given above, $3 \mid N$ if and only if 3 divides the digital root of N.

All this follows from the fact that the remainder is always 1, when any power of 10 is divided by 3.

Here is a program to compute the digital root of a (base 10) number:

```
  5  X = 0
 10  PRINT  "[CLR/HOME]"
 20  PRINT  "PROGRAM COMPUTES THE DIGITAL"
 30  PRINT  "ROOT OF N. PLEASE ENTER N."
 40  INPUT N$
 50  N1$=N$
 60  L=LEN(N$)
 70  FOR I=1 TO L
 80  X=X+VAL(MID$(N$,I,1))
 90  NEXT I
100  IF X<10 THEN 200
110  N$=STR$(X)
120  X=0
130  GO TO 60
200  PRINT X;" IS THE DIGITAL ROOT OF ";N1$
210  END
```

Let's discuss how the program works. Before we can compute the digital root of a number, we need to tell the computer how to identify its digits. (Remember, in its own calculations, the computer uses base 2.) One approach would be to divide the number, N, by 10 getting a quotient, Q1, and a remainder, A. Then A is the 1's digit in N. Here is a routine to find Q1 and A. (Don't ENTER it. This is only an illustration; it is not the algorithm we have chosen to implement just now.)

```
500  Q1=INT(N/10)
510  A=N−10*Q1
```

We could then divide Q1 by 10, obtaining a quotient Q2 and a remainder B. Then B is the 10's digit in the base 10 expression for N. Etc.

The program LISTed above implements another algorithm. Beginning with line 40, we instruct the computer to treat the INPUT as a string variable. We are going to isolate the individual digits of N by means of some new string variable commands.

The new BASIC code word LEN calls for the LENgth of the string; that is, LEN(N$) is the number of symbols in N$. We're interested in LEN(N$) because it is the number of digits in the number N.

Before continuing, experiment with LEN. Try some of these statements:

```
PRINT LEN("ABCDE")
PRINT LEN("123")
PRINT LEN("8H56T;2")
PRINT LEN(1286)
```

Observe that the computer doesn't know how to interpret the LENgth of a number. A TYPE MISMATCH ERROR, or some such message, results because the computer expected to see a string and found something else. It is confused.

The BASIC code word MID$ forms a new string by taking just part of an existing string. Experiment with these statements:

```
X$="12ABC3456"
PRINT MID$(X$,3,1)
PRINT MID$(X$,3,2)
PRINT MID$(X$,3,3)
PRINT MID$(X$,6,4)
PRINT MID$("QWERTY",2,2)
```

There are three "arguments" in the MID$ instruction. The first is a string, while the second and third are numeric. The second argument is the starting point for the portion of the string to be extracted, and the third is the LENgth of the resulting portion.

Lines 70–90 of the program (above) compute the sum of the digits of N. VAL(MID$(N$,I,1)) is the numeric VALue of the I-th symbol in the string N$, that is, of the I-th digit of N. If X<10, it is the digital root of N, and we PRINT it out in line 200. Otherwise, we cycle through with the new "number" N$ = STR$(X).

Let's return to the discussion of prime numbers. Recall our improved strategy for deciding whether a given number n is prime or composite: "If n is composite, then it has a prime divisor $p \leq \sqrt{n}$." Thus, if we know the primes $\leq \sqrt{n}$, we can systematically check whether one of them divides n and eventually prove either that n is prime, or find one of its prime divisors. This is the algorithm we might use for hand computation; and, so, it is the algorithm we "taught" the computer by means of the program in Section 6.1.

It is somewhat ironic that there is another algorithm which is both easier to ENTER, and easier (faster) for the computer as well. We went to a great deal of trouble so that the computer wouldn't have to take the time to divide, say, 2047 by composite numbers. The trouble is that one of the things computers do fastest is arithmetic. We managed, inadvertently, to slow the machine down by making it search for the next number in its list of DATA. It would have been faster to let the machine churn through a few unnecessary composite divisors.[†]

Here is a compromise program for determining whether N is prime. The "compromise" is to make use of the fact that primes > 2 are odd.

```
10  INPUT "WHAT IS THE NUMBER";N
20  IF N<10 THEN 100
30  IF N/2=INT(N/2) THEN 100
40  K=INT(SQR(N))
50  IF K/2=INT(K/2) THEN K=K-1
60  FOR I=3 TO K STEP 2
70  IF N/I<>INT(N/I) THEN 80
75  PRINT I;"DIVIDES";N:GO TO 110
80  NEXT I
90  PRINT N;"IS PRIME.":GO TO 110
100 PRINT "DON'T WASTE MY TIME."
110 END
```

RUN the program for N = 3,613,801, and see how long the computer takes to resolve the primality of N.

Exercises (6.2)

1. Determine which of the following numbers are prime. Find the smallest prime factor of each composite number.

   a.   65,537   b.   131,071   c.   524,287   d. 912,673
   e. 1,113,121   f. 3,648,091   g. 8,388,607

[†]Fastest of all would be a BASIC instruction like FOR I=1 TO K STEP [PRIME]. The trouble is that each BASIC statement constitutes a "machine language" program either built in to the machine at the factory, or ENTERed by disk when the machine is turned on. Previous programmers have simplified your job by instructing the computer how to interpret BASIC commands. Unfortunately, no one knows how to write a program for STEP [PRIME]. Put another way, no one knows the "formula" for prime numbers, that is, a function $p(x)$ such that $p(n)$ is the n-th prime.

2. What happens if the 3rd numeric variable in MID$ is too big? ENTER and record the output for each of these:
   a. PRINT MID$("ABCD",1,5)
   b. PRINT MID$("12345",4,3)

3. This is an exercise for you, not the computer. What responses would the computer give to:
   a. PRINT MID$("ABCDEF",2,3)
   b. PRINT MID$("HIGH SCHOOL",3,5)
   c. PRINT MID$("$1,000",2,2)
   d. PRINT MID$($1,000,000)
   e. PRINT MID$("R2D2 & C–3PO",4,5)
   f. PRINT MID$("R2D2 & C–3PO",8,6)

4. There are two variations on MID$ that are useful if you want a portion of a string starting from the left or ending at the right. What response does the computer give to:
   a. PRINT LEFT$("YES",1)
   b. PRINT LEFT$("12345",3)
   c. PRINT RIGHT$("FOREVER",4)
   d. PRINT RIGHT$("12345",3)

5. What is the difference between MID$(A$,1,n) and LEFT$(A$,n)?

6. Express RIGHT$(A$,n) in terms of MID$. (*Hint:* Use LEN.)

7. Test the expression you devised in Exercise 6 on the computer for several examples. Record the examples and the computer's responses.

8. If both N and N+2 are prime, they are called a prime pair. Thus, 11 and 13 constitute a prime pair as do 5441 and 5443. Find all the prime pairs less than 100. (*Hint:* Section 6.1, Exercise 2.)

9. If N, N+2, and N+4 are all prime, they are called a prime triple. Prove that 3, 5, and 7 is the only prime triple.

10. Prove that 9 | N if and only if the digital root of N (base 10) is 9.

11. Prove that 7 | N if and only if the digital root of N (base 8) is 7.

12. Write a program to INPUT N and PRINT the prime pairs less than N. (See Exercise 8.) In addition, PRINT a running count of the prime pairs as they're produced.

13. Make a list of the 35 prime pairs less than 1000. (*Hint:* See Exercise 12.)

14. How many prime pairs are there less than 5000?

15. Let $N = 2*3*5*7*11*13 + 1$. What is the remainder when N is divided by:

    a. 2   b. 3   c. 5   d. 7   e. 11   f. 13

    (*Hint:* It shouldn't be necessary to "multiply out.") Is N prime?

16. In 1742, Christain Goldbach noticed that every even number greater than 2 seems to be the sum of two primes. (For example, $4 = 2 + 2, 6 = 3 + 3, 8 = 5 + 3, 10 = 7 + 3 = 5 + 5, \ldots$.) Being unable either to prove or disprove his observation, Goldbach wrote to the greatest mathematician of his time, Leonhard Euler (1707–1793). Euler became convinced that Goldbach's observation is always true, but he couldn't prove it. To this day, no one is sure whether "Goldbach's Conjecture" is true or false. Write every even number from 12 to 30 as a sum of two primes in all possible ways.

6.3 COMPUTATIONAL COMPLEXITY

As we know, the solution of a problem by computer requires a sequence of precise instructions, an algorithm. While algorithms occur in some of the earliest of human records, it is the rise of computers and computer science that has led to the widespread study of algorithms for their own sake.

Above all, there can be no ambiguity in a computer algorithm. Each step of the execution must be uniquely defined and may depend only on the INPUTs, the previous steps, and the internal capabilities of the machine. But, suppose two algorithms are available to do the same job. Assuming both are correct, how does one choose the best one?

Generally, there are two main considerations: (1) How long does it take to RUN? and (2) How much memory space is needed? These are the components of what has come to be called the *complexity* of the algorithm. Giving precise quantitative estimates for complexity is difficult because so may factors are involved. To take just one example, should we base our quantitative estimates on the best case, the worst case, or the average case? Consider the example of whether N is prime.

Apart from small numbers, the easiest is the case in which N is even. The first prime we try will divide N. The worst case is that in which N is prime, or the square of a prime. In both of these sitautions, we have to check through the maximum number of possibilities.

While the space requirements of an algorithm are every bit as important to the advanced programmer as time, we are going to ignore memory considerations and restrict our attention to time. Execution time is a function (among other things) of the size of the INPUT. Call this n. Since it is usually hopeless to give exact estimates for time in terms of n, one usually settles for upper bounds. Such bounds are often written using the "big oh" notation.

Suppose $f(x)$ is a function. We say the time to RUN an algorithm is of order $f(n)$, and write $O(f(n))$, if there is a positive constant C such that the actual time to RUN is $\leq C\,|\,f(n)\,|$, for all but finitely many n.

Consider the program at the end of the last section (for determining if N is prime). What should n be? After all, only one number is being entered. Should n be 1? No. In this case, the proper measure for the size of the INPUT is $n = $ N. Since we must do at most $\sqrt{n}$ divisions, the time to RUN the algorithm is proportional to $\sqrt{n}$. In other words, there is a constant C such that the time is $\leq C\sqrt{n}$. This algorithm is $O(\sqrt{n})$.[†] But, wait a minute. The algorithm we finally settled on was a "compromise." In a single step, we eliminated half the divisors (the even ones)! Thus, we really have to check fewer than $\sqrt{n}/2$ divisors. But, the factor $1/2$ winds up affecting only the constant, C. That is to say, $C*(\sqrt{n}/2) = (C/2)*\sqrt{n}$ is just a constant times $\sqrt{n}$. As we don't care what C is anyway, $O(\sqrt{n}) = O(\sqrt{n}/2)$. In other words, even the so-called quantitative estimate, $O(\sqrt{n})$, is extremely rough. It is a very large ballpark estimate.

An $O(\sqrt{n})$ algorithm is better than an $O(n)$ algorithm, which is better than an $O(n^2)$ algorithm. In a qualitative sense, algorithms are "good" if there is a "polynomial" function $f(x)$ such that the algorithm is $O(f(n))$.[‡] An example of a "bad" algorithm is, for example, one that is $O(2^n)$. "Exponential" algorithms are bad.

[†] The constant C depends on the machine. The faster the machine, the smaller the constant.

[‡] See Section 7.2.

Let's return to the problem of determining whether N is prime. Maybe we can squeeze a little more out of our "compromise" and improve on $O(\sqrt{n})$. The initial compromise consisted of eliminating every other number, the even ones, those which are divisible by 2. Why not also eliminate multiples of 3? Indeed, consider the positive integers starting with 2:

	2	3	4	5	6	7	8	9	10
11	12	13	14	15	16	17	18	19	20
21	22	23	24	25	26	27	28	29	30
31	32	33	34	35	36	37	38	39	40
...									

Circle 2 and cross off every second number after that.

	②	3	~~4~~	5	~~6~~	7	~~8~~	9	~~10~~
11	~~12~~	13	~~14~~	15	~~16~~	17	~~18~~	19	~~20~~
21	~~22~~	23	~~24~~	25	~~26~~	27	~~28~~	29	~~30~~
31	~~32~~	33	~~34~~	35	~~36~~	37	~~38~~	39	~~40~~
...									

Now circle 3, and cross off every third number after that.

	②	③	~~4~~	5	~~6~~	7	~~8~~	~~9~~	~~10~~
11	~~12~~	13	~~14~~	~~15~~	~~16~~	17	~~18~~	19	~~20~~
~~21~~	~~22~~	23	~~24~~	25	~~26~~	~~27~~	~~28~~	29	~~30~~
31	~~32~~	~~33~~	~~34~~	35	~~36~~	37	~~38~~	~~39~~	~~40~~
...									

The next number after 3 which has not been crossed off is 5. Since it hasn't been crossed off, it is not a multiple of a smaller prime (that is, 2 or 3). Thus (we can reason), 5 must be prime. Circle it, and cross off every fifth number thereafter.

②	③	4	⑤	~~6~~	7	8	~~9~~	~~10~~
11	~~12~~	13	~~14~~	~~15~~	~~16~~	17	~~18~~	19	~~20~~
~~21~~	22	23	~~24~~	~~25~~	26	~~27~~	28	29	~~30~~
31	32	~~33~~	34	~~35~~	~~36~~	37	38	~~39~~	~~40~~

. . .

At this point, every number that has not been crossed off must be prime. The smallest prime dividing any number ≤ 40 is at most $\sqrt{40}$, hence at most 5. (The next prime after 5—namely, 7—is larger than $\sqrt{40}$. If the list had extended to 50, then 49 would be the only composite number remaining to be crossed off. This algorithm for finding primes was invented by the Greek mathematician Eratosthenes (ca. 275–195 B.C.).[†] It is generally known as the "Sieve of Eratosthenes."

What if we incorporate the entire Sieve of Eratosthenes into a primality program? Then, it turns out, we will have essentially reverted to the original algorithm of Section 6.1! In other words, at some point, we must start adding time, making things harder and longer, not faster! It is considerations such as these which move us beyond the realm of computer programming into that of computer science.

Suppose n is a positive integer greater than 1. Then, according to our definitions, n is either prime or composite. If n is composite, it has a proper prime divisor, call it p_1. Then $n = p_1*q_1$. But, what about q_1? Since p_1 is a proper divisor, $q_1 > 1$. Thus, q_1 is either prime or composite. If q_1 is a prime, then we have expressed n as a product of primes. If q_1 is composite, it has a smallest prime divisor, call it p_2. So, $q_1 = p_2*q_2$, and $n = p_1*p_2*q_2$. If q_2 is prime, then n is a product of primes. If not, then q_2 is composite, and it has a smallest prime divisor. This process cannot continue forever because the successive quotients, q_1, $q_2, \ldots$, keep getting smaller. Eventually the process must stop. Now, the only way it can stop is if one of the quotients is prime, that is, if n is equal to a product of primes. We have arrived at a result that helps to explain why primes are so important:

[†] In spite of some myths, neither Columbus nor Magellan "discovered" that the Earth is "round." Eratosthenes not only knew that the Earth is a sphere, he also managed to give an accurate estimate of its circumference!

THEOREM:

Every integer greater than 1 is either prime or a product of primes.

If we are willing to think of 3, say, as a "product" of primes—namely, just one prime—then the Theorem can be simplified as follows: Every integer greater than 1 is a product of primes.

Notice that the argument leading up to the theorem is both a proof and an algorithm. Not only does it prove that every integer greater than 1 is a product of primes, it describes how to find the primes.

Let's work out an example. Suppose $n = 360$. The smallest prime dividing 360 is 2, so $p_1 = 2$ and $q_1 = 180$. The smallest prime dividing 180 is $p_2 = 2$, so $q_2 = 90$. Similarly, $p_3 = 2$ and $q_3 = 45$; $p_4 = 3$ and $q_4 = 15$; $p_5 = 3$ and $q_5 = 5$. Because q_5 is prime, the process stops. Assembling the pieces,

$$360 = 2*180$$

$$= 2*2*90$$

$$= 2*2*2*45$$

$$= 2*2*2*3*15$$

$$= 2*2*2*3*3*5$$

Alternatively,

$$360 = 2^3*3^2*5$$

In addition to avoiding ambiguity, an algorithm must be finite. An infinite process is ordinarily no good even to the fastest computer. What guarantees that the above process is finite is that

$$n > q_1 > q_2 > q_3 > \ldots > 1$$

In particular, the algorithm must terminate before n steps. Indeed, we can say a bit more. Since $n = p_1*q_1$, where p_1 is prime, $q_1 \leq n/2$. In general, at each stage, the quotient is (at least) cut in half. It follows that the process must end by the m-th step, where m is the largest integer for which $2^m \leq n$. If 2^m were equal to n, then $m = \log_2(n)$. Thus, the absolute worst case is that the algorithm will involve $\log_2(n)$ steps. Because finding the smallest prime factor of the i-th quotient is $O(\sqrt{q_i}) < O(\sqrt{n})$, the complete algorithm is (at worst) $O(\sqrt{n}*\log_2(n))$.

Exercises   (6.3)

1. By hand, without recourse to a computer or calculator, write each of the following as a product of primes.
   a. 243   b. 343   c. 344   d. 345   e. 567

2. It can be shown that the best algorithms for sorting n numbers into numerical order are $O(n*\ln(n))$. Which is better, $O(n*\ln(n))$ or $O(n^2)$?

3. Which is better, $O(\sqrt{n})$ or $O(\ln(n))$? (*Hint:* RUN this program.)

```
10  FOR N=1 TO 100
20  PRINT SQR(N);LOG(N)
30  FOR I=1 TO 200
40  NEXT I
50  NEXT N
60  END
```

4. Which is better, $O(\log_2(n))$ or $O(\ln(n))$? (*Hint:* See Section 5.4, Equation (5.12).)

5. For large values of n, say 1 million, n^3 is 1 million times larger than n^2. Thus, n^2 is insignificant when compared to n^3. In particular, $O(n^3 + n^2) = O(n^3)$. More generally, if $p(x)$ is a polynomial of degree k, then $O(p(n)) = O(n^k)$. Use this observation to simplify
   a. $O(n^4 + n)$ b. $O(n^2 + 5n)$ c. $O(n^2 + 1000n + 1,000,000)$

6. Write a program to INPUT a number N and output the same number expressed as a product of primes.

7. Express each of the following as a product of primes:
   a. 12345   b. 123456   c. 54321   d. 31416   e. 314159

8. Demonstrate the Sieve of Eratosthenes for the first 100 numbers. (*Hint:* It isn't necessary to cross off every 11th number. Why?)

9. Pierre de Fermat (1601–1665) proved that for any prime p, and for any positive integer n, $p \mid (n^p - n)$. Confirm this result by finding the quotient in each of the following instances.
   a. $n = 17$ and $p = 2$ b. $n = 2$ and $p = 17$
   c. $n = 6$ and $p = 7$ d. $n = 8$ and $p = 7$

10. Prove Fermat's Theorem (Exercise 9) for the special case that $p = 2$.

11. Write a computer program to confirm Fermat's Theorem (Exercise 9) for $2 \leqslant n \leqslant 10$ and $p = 3, 5,$ or 7.

6.4 GREATEST COMMON DIVISORS

If m and n are two positive integers, then k is said to be a common divisor of m and n if $k \mid m$ and $k \mid n$. The positive integer d is the *greatest common divisor* (or GCD) of m and n if (1) $d \mid m$ and $d \mid n$; and (2) $k \mid d$ for any common divisor k of m and n. For example, 2 is a common divisor of 8 and 12, but 4 is their greatest common divisor. We will use the notation (m,n) to denote the greatest common divisor of m and n. Thus, $(8,12) = 4$, $(2,3) = 1$, and $(30,42) = 6$.[†]

For small numbers m and n, one can usually determine the GCD by trial and error. This method is inefficient, however, for large m and n. What, for example, is the GCD of 248653 and 246031?[‡] The following lemma[§] is the heart of an algorithm for finding GCD's.

LEMMA:

If $m + nk = r$, then $(m,n) = (n,r)$.

Proof. Let d_1 denote (m,n) and d_2 denote (n,r). Since $d_1 \mid m$ and $d_1 \mid n$, there are quotients q_1 and q_2 such that $m = d_1 * q_1$ and $n = d_1 * q_2$. Then

$$m + n*k = d_1*q_1 + d_1*q_2*k$$

$$= d_1*(q_1 + q_2*k)$$

In other words, $d_1 \mid (m + nk)$, so $d_1 \mid r$.

Among all the things we know are the facts that $d_1 \mid n$ and (now) that $d_1 \mid r$. It follows from the definition that d_1 divides the GCD of n and r, that is, $d_1 \mid d_2$.

[†]More generally, one can discuss common divisors of any two nonzero integers. Without the restriction that the GCD be positive, there would be two greatest common divisors, a positive one, d, and a *negative one*, $-d$.

[‡]Of course, $(248653, 246031)$ is a *name* for the answer as opposed to the answer itself.

[§]"Lemma" is a word used to indicate a theorem that is of interest primarily for its usefulness in obtaining other results.

Next, we'll show that $d_2 \mid d_1$. Rewrite the equation as $m = -nk + r$. Since $d_2 \mid n$ and $d_2 \mid r$, we can write $n = d_2*q_3$ and $r = d_2*q_4$. Then

$$m = -n*k + r$$

$$= -(d_2*q_3)*k + d_2*q_4$$

$$= d_2*(q_4 - q_3*k).$$

It follows that $d_2 \mid m$. So, we have (among other things) $d_2 \mid m$ and $d_2 \mid n$. So, d_2 is a common divisor of m and n. It must, therefore, divide d_1, the GCD of m and n.

So, $d_1 \mid d_2$ and $d_2 \mid d_1$. There are quotients q and q' such that $d_2 = q*d_1$ and $d_1 = q'*d_2$. Putting these together, we see that $d_2 = q*q'*d_2$. If we cancel d_2, we obtain $1 = q*q'$. Since both q and q' are positive integers, they must both be 1, that is, $d_1 = d_2$, or $(m,n) = (n,r)$. Q.E.D.

Let's see how the lemma yields an algorithm for producing the GCD of m and n. We might as well assume that m is the larger of the two integers. If $n \mid m$, then $(m,n) = n$. If n doesn't (exactly) divide m, we can still "divide" n into m, and obtain a quotient q and a remainder r. Thus,

$$m = n*q + r \tag{6.3}$$

(or $m + n(-q) = r$). It follows from the lemma that $(m,n) = (n,r)$. So what? Why trade one GCD for another?

When you divide one whole number by another, say m by n, how do you know when to stop dividing? How do you know when you've reached the remainder? You know it's time to stop when you come to a number, r, which is less than n, the number you are dividing by. So, in Equation (6.3), $r<n$. Since we're assuming $n<m$, the reason to trade (m,n) for (n,r) is that $r<m$. We get to deal with smaller numbers. Consider the following process:

$$m = n*q_1 + r_1$$

$$n = r_1*q_2 + r_2$$

$$r_1 = r_2*q_3 + r_3$$

$$r_2 = r_3*q_4 + r_4$$

$$. . .$$

Since $n > r_1 > r_2 > r_3 > r_4 \cdots \geq 0$, the process can't go on forever. It is a finite process. It must stop sometime; but, how does it stop? It stops when one of the remainders is zero. Suppose, for example, that $r_4 = 0$. Then $r_2 = r_3 * q_4$. So, r_3 is the GCD of r_2 and r_3. But, using the lemma on the previous equation, $(r_1, r_2) = (r_2, r_3)$. Thus, r_3 is the GCD of r_1 and r_2. But then, using the lemma on the next previous equation— namely, $n = r_1 * q_2 + r_2$—we can conclude that $(n, r_1) = (r_1, r_2)$. So, r_3 is also the GCD of n and r_1. We have already argued that $(m, n) = (n, r_1)$. Hence, r_3 is the GCD of m and n.

In summary, the algorithm for finding the GCD of two numbers is this:

1. Divide the larger by the smaller, obtaining a quotient and remainder.

2. If the remainder is 0, the smaller is the GCD.

3. Otherwise, replace the larger with the smaller. Replace the smaller with the remainder. Go to step 1.

Let's work out a numerical example: What is the GCD of 690 and 3030?

$$3030 = 690 * 4 + 270$$

$$690 = 270 * 2 + 150$$

$$270 = 150 * 1 + 120$$

$$150 = 120 * 1 + 30$$

$$120 = 30 * 4$$

We stop when we reach a step in which the remainder is zero. (Notice that the remainders do, in fact, strictly decrease from one step to the next.) To review the logic behind the algorithm for this example,

$$30 = (30, 120)$$

$$= (120, 150)$$

$$= (150, 270)$$

$$= (270, 690)$$

$$= (690, 3030)$$

Of course, you needn't "review the logic" (except, perhaps, in your head) every time you compute the GCD of two numbers. The answer is, simply, 30.

We say two integers m and n are *relatively prime* if $(m,n) = 1$. In other words, m and n are relatively prime if 1 is the only positive integer that (exactly) divides them both. For example, 10 and 21 are relatively prime. (Note that neither 10 nor 21 is itself prime. Of course, if either m or n is prime, then m and n are relatively prime.)

Exercises (6.4)

1. Find the following greatest common divisors:
   a. (2,8)   b. (6,15)   c. (9,15)   d. (10,15)

2. Explain why $(m,n) = (n,m)$ for any pair of positive integers m and n.

3. Explain why two positive integers can have only one GCD.

4. Write a computer program to: (1) INPUT two positive integers, M and N; (2) PRINT their GCD; and (3) along the way, PRINT each equation in the algorithm for producing (M,N), that is, each equation of the form $r_{(k-1)} = r_k * q_{(k+1)} + r_{(k+1)}$.

5. Find the following greatest common divisors:
   a. (3105,345)   b. (1377,6531)   c. (8910,5670)
   d. (2243,9560)   e. (6240,9216)   f. (10080,12345)

6. What is (248653,246031)?

7. If $(m,n) = d$, prove that $(m/d,n/d) = 1$.

8. Suppose $(m,n) = d$. Let $k = mn/d$. Prove that $m \mid k$ and $n \mid k$, that is, that k is a common *multiple* of m and n.

9. If n is a positive integer, define $f(n)$ to be the number of positive integers less than n that are relatively prime to n. Then f is known as the "Euler phi function."[†] For example, $f(8) = 4$. The positive integers less than 8 and relatively prime to 8 are 1, 3, 5, and 7. Find $f(n)$ for $n = 2, 3, 4, \ldots , 20$.

[†] Euler chose the notation $\phi(n)$ for this function. The use of ϕ here is unrelated to its use as the "golden ratio" in Chapter 2.

10. Let f be the Euler phi function. (See Exercise 9.) Show that $f(n) \leq n-1$ with equality if and only if $n = 1$ or n is prime.

11. Each of the following numbers, n, has the property that if $1 < m < n$, and $(m,n) = 1$, then m is prime. Confirm that this is so, and compute $f(n)$ in each case. (See Exercise 9.)
    a. $n = 3$ b. $n = 4$ c. $n = 6$ d. $n = 8$
    e. $n = 12$ f. $n = 18$ g. $n = 24$ h. $n = 30$

12. Can you see a pattern in the following sequence?

$$3, 4, 6, 8, 12, 18, 24, 30, \ldots.$$

(See Exercise 11.) What is the next number n, larger than 30, such that $1 < m < n$, and $(m,n) = 1$, only if m is prime?

13. Confirm that:
    a. $1 = 2 - 1$
    b. $1 + 2 = 2^2 - 1$
    c. $1 + 2 + 2^2 = 2^3 - 1$
    d. $1 + 2 + 2^2 + 2^3 = 2^4 - 1$
    e. $1 + 2 + 2^2 + 2^3 + 2^4 = 2^5 - 1$

14. Prove that $1 + 2 + 2^2 + \cdots + 2^{n-1} = 2^n - 1$. (*Hint:* See Section 1.5.)

15. Show that the GCD algorithm given in the text is $O(\log(n))$, where the "n" mentioned here is the larger of the two numbers. (*Hint:* To avoid confusion arising from two uses of the letter "n," let's talk in terms of the GCD of a and b. If a is the larger of the two, then $r_1 < a/2$. Why? $r_2 < b/2$ and $r_2 < r_1$. Why? $r_3 < r_1/2 < a/4$. Why? ... Argue that the process must terminate in at most $2*\log_2(a)$ steps.)

16. Define the GCD of three positive numbers, say (m,n,k). Compute
    a. $(8,10,12)$ b. $(18,48,132)$ c. $(28,70,350)$
    d. $(64,16,1024)$ e. $(2,3,6)$

6.5 THE FUNDAMENTAL THEOREM

Sometimes, we want to RUN a program repeatedly. Maybe we have several numbers we want to factor, or several pairs of numbers whose GCD we wish to compute. One way to avoid having to ENTER the

RUN command each time is to put an instruction at the end of the program to GO TO the beginning. In some instances, this can lead to trouble. The RUN instruction automatically clears all variable memory locations so that we are free to assume these variables have the default value of zero. GO TO the beginning leaves these memory locations unaltered. The BASIC command CLR has the effect of CLeaRing all variables, arrays,[†] FOR . . . NEXT loops, and GOSUB addresses (to which program execution might RETURN). Thus, if you plan to loop through a program several times, you might wish to begin it with a line like

```
10 CLR
```

The disadvantages of ending the program with a GO TO the beginning command is that it is awkward to terminate execution when you are finished. To be sure, the program interrupt procedure is available, but here is another option. Instead of ending with GO TO the beginning, consider this:

```
1000 PRINT "DO YOU WANT TO LOOP THROUGH"
1010 INPUT "AGAIN";A$
```

The idea is that you should GO TO 10 only when A$ is some version of "yes." Here is an option for the next line that permits several variations on "yes":

```
1020 IF LEFT$(A$,1)="Y" THEN 10
1030 END
```

This allows you the flexibility to INPUT YES, YESSIREE, YUP, or just Y at line 1010. Try a simple experiment or two; for example,

```
10 CLR
20 DIM N(15)
30 FOR I=1 TO 5
40 A=A+I
50 PRINT A
60 NEXT I
70 INPUT "AGAIN";A$
80 IF LEFT$(A$,1)<>"N" THEN 10
90 END
```

[†] Including DIM statements so that the corresponding program line can be executed again without a REDIM'D ARRAY ERROR.

(Note that the computer has no trouble distinguishing A from $A\$$.) Loop through a couple of times, and then try it with this substitution for line 10:

```
10 REM
```

On the second pass, you should get a REDIM'D ARRAY ERROR. Eliminate line 20 and RUN again, looping through several times. With CLR gone, you should notice a different response from the first RUN.

Let's return to the discussion of greatest common divisors. Consider the GCD of 138 and 54,

$$138 = 54*2 + 30$$

$$54 = 30*1 + 24$$

$$30 = 24*1 + 6$$

$$24 = 6*4$$

Thus, $(138,54) = 6$. Let's see how to squeeze a little more out of the process.

It turns out always to be possible to express (m,n) in the form $mx + ny$, for some integers x and y. (One of x and y will be positive and the other negative.) Not only that, but we can actually obtain x and y by working backwards in the process above. From the next to the last equation in the example above,

$$6 = 30 - 24*1 = 30 - 24 \tag{6.4}$$

Solving for 24 in the second equation in the example and substituting in Equation (6.4), we obtain

$$6 = 30 - (54 - 30) = 30*2 - 54 \tag{6.5}$$

Finally, solving for 30 in the first equation and substituting the result in Equation (6.5), we get

$$6 = (138 - 54*2)*2 - 54 = 138*2 - 54*5$$

So, $6 = 138x + 54y$, where $x = 2$ and $y = -5$. (Notice that we don't want to do any actual arithmetic! In Equation (6.5), for example, we do not want to multiply 30 and 2 to get 60.)

Lemma:

Suppose p is a prime and $p \mid mn$. Then $p \mid m$ or $p \mid n$.

Proof. If $p \mid m$, we are finished. If p does not divide m, then p and m must be relatively prime, that is, $(m,p) = 1$. The GCD of m and p is 1. It follows from our little argument above, there are integers x and y such that

$$mx + py = 1.$$

Multiply both sides of this equation by n. The result is

$$mnx + pny = n$$

Since $p \mid mn$, there is an integer q such that $mn = pq$. Substituting this in the last equation, we get

$$pqx + pny = n$$

$$p(qx + ny) = n$$

In other words, $p \mid n$. Q.E.D.

We can now improve on the theorem in Section 6.3. Not only is every integer $n>1$ a product of primes, there is essentially only one way to factor n as such a product. This refined result is known as the *Fundamental Theorem of Arithmetic*:

THEOREM:

Every integer greater than 1 can be factored, in exactly one way, as a product of primes.

Proof. We already know that every integer $n>1$ can be factored as a product of primes. It only remains to show that there is a unique way to do it. Suppose, to the contrary, that there were two ways to do it, maybe

$$n = p_1 * p_2 * \ldots$$

and

$$n = q_1 * q_2 * \ldots$$

Then

$$p_1 * p_2 * \ldots = q_1 * q_2 * \ldots$$

If these really are two different factorizations (except for the order in which the factors are written), then we can cancel the common primes and still have something left, say

$$p_1*p_2*\ldots*p_k = q_1*q_2*\ldots*q_r, \qquad (6.6)$$

where all the p's and q's are primes, and none of the p's is equal to any of the q's. We will soon see that this is impossible. Clearly, p_1 divides the left-hand side of Equation (6.6). Hence, it divides the right-hand side. So, p_1 divides $q_1*(q_2*\ldots*q_r)$. By the lemma, p_1 either divides q_1 or p_1 divides $(q_2*\ldots*q_r)$. Certainly the prime p_1 cannot divide the prime q_1 without being equal to it. Since we're operating under the assumption that none of the p's, in particular p_1, can be found among the q's, $p_1 \neq q_1$. It must be that p_1 divides $q_2*(q_3*\ldots*q_r)$.

By another application of the same argument, p_1 must divide $q_3*\ldots*q_r$ $= q_3*(q_4*\ldots*q_r)$. Eventually, we must reach the conclusion that $p_1 \mid q_r$. But that, also, is impossible. The whole situation that arises from Equation (6.6) is untenable . . . unless there is no p_1. The only way out of the difficulty is that there are no primes left over after cancellation, that (6.6) is really just $1 = 1$. Q.E.D.

One application of the Fundamental Theorem of Arithmetic is that a number like 53*59 could never equal something like 47*61. We don't have to multiply anything out to convince ourselves. Since 47, 53, 59, and 61 are all primes, if 53*59 were to equal 47*61, then their common product would have two different prime factorizations, contradicting the theorem.

Exercises (6.5)

1. Is it possible that a product like 24*75 could equal one like 30*60? Justify your answer.

2. Factor each number as a product of primes:
   a. 6   b. 28   c. 220   d. 284   e. 496   f. 8128

3. By hand, without using the computer, find all the positive divisors (not just the prime divisors) of
   a. 6   b. 28   c. 220   d. 284   e. 496   f. 8128
   (*Hint:* Use the Fundamental Theorem of Arithmetic. The divisors of a positive integer will be 1, all the prime divisors, all products of prime divisors taken two at a time, all products of prime divisors taken three at a time, . . .).

4. Write a computer program to INPUT N and output all of its divisors.

5. Discuss any differences between the algorithm you used yourself in Exercise 3, and the one you are making the computer use in Exercise 4.

6. Find all the divisors (not just the prime divisors) of
   a. 12496   b. 14288   c. 15472   d. 14536   e. 14264

7. If n is a positive integer, let $S(n)$ denote the sum of all of its positive divisors except for n itself. Show that:
   a. $S(12496) = 14288$
   b. $S(14288) = 15472$
   c. $S(15472) = 14536$
   d. $S(14536) = 14264$
   e. $S(14264) = 12496$

8. Suppose a and b are positive integers. If $a \mid m$ and $b \mid m$, then m is a common multiple of a and b. If l is a common multiple of a and b which divides any other common multiple, then l is the *least common multiple* (or LCM) of a and b. Sometimes the LCM of a and b is denoted $[a,b]$. Find:
   a. $[6,15]$ b. $[30,42]$ c. $[5,6]$ d. $[5,10]$

9. Prove that $ab/(a,b)$ is the least common multiple of a and b. (See Exercise 8.)

10. Write a program to INPUT A,B and output their LCM. (See Exercise 8.)

11. Find
    a. $[3105,345]$ b. $[391,437]$ c. $[8910,5670]$

6.6 PERFECT NUMBERS

Two of the things that come to mind when "mathematics" is mentioned are numbers, and a strict adherence to logical thinking. These two concepts have not always been so closely related. Indeed, even today, there is a substantial minority that ascribes mystical properties to certain numbers. There are, for instance, many numerological allusions in the biblical literature, especially in the Book of Revelations. We are told, for example, that the "number of the beast" is 666. The ancients were particularly attracted, to the so-called "perfect" numbers. A number, n, is said to be *perfect* if it is equal to the sum of all its divisors except for n itself. The smallest perfect number is:

$$6 = 1 + 2 + 3$$

In *The City of God*, St. Augustine argued that God, who could have made the universe in an instant, chose to labor for six days in order to emphasize the perfection of his creation.

The next perfect number is:

$$28 = 14 + 7 + 4 + 2 + 1,$$

which happens to correspond to the number of days in a lunar month. After 28 comes 496, and then 8128, the largest perfect number known before 300 B.C.

Euclid, who flourished about 300, B.C.,[†] is remembered today primarily as a geometer. The ninth book of his *Elements*, however, contains a good deal of what he might have called arithmetic, what we now know as number theory. He proved, for one thing, that there are infinitely many primes. Unlike the atoms, of which only 92 occur in nature, the building blocks of the integers are without bound. In the same ninth book, Euclid established a way to construct perfect numbers.

THEOREM (Euclid):

If p is a prime and if $2^p - 1$ is prime, then $n = 2^{p-1}(2^p - 1)$ is perfect.

Proof: By the Fundamental Theorem of Arithmetic, the divisors of n are precisely

1	2	2^2	$\ldots$	2^{p-1}
(2^p-1)	$2(2^p-1)$	$2^2(2^p-1)$	$\ldots$	$2^{p-1}(2^p-1)$

[†] In 300 B.C., Alexander the Great had been dead 23 years. Ptolemy Soter ruled in Egypt, and the Greek general Pyrrhus had begun to resist the growing power of Rome. (Rome's first war with Carthage was still 64 years in the future.)

For any k, $2^k(2^p - 1) = 2^{k+p} - 2^k$. So, in the sum, s, of *all* the divisors of n, there is some cancellation. When the dust settles, we see

$$s = 2^p + 2^{p+1} + 2^{p+2} + \cdots + 2^{2p-1}$$

$$= 2^p(1 + 2 + 2^2 + \cdots + 2^{p-1})$$

$$= 2^p(2^p - 1)$$

$$= 2(2^{p-1})(2^p - 1)$$

$$= 2n ,$$

where the third step follows from Section 6.4, Exercise 14. Q.E.D.

Let's try a few examples. First, for any positive integer, k, let

$$M_k = 2^k - 1$$

For $p = 2$, $M_2 = 2^2 - 1 = 3$. Since 3 is prime, the hypothesis of the theorem is operative. We conclude that $2^{2-1} * M_2 = 2*3 = 6$ is perfect.

Clearly, we didn't need Euclid's Theorem to tell us that 6 is perfect. On the other hand, it is interesting to note that the first perfect number arising from the theorem is the smallest one. For $p = 3$, $M_3 = 2^3 - 1 = 7$, also prime. So $2^{3-1} * 7 = 4*7 = 28$ is perfect. Corresponding to the third prime, 5, is $M_5 = 2^5 - 1 = 31$. So, the next perfect number that can be deduced from the theorem is the third perfect number, $16*31 = 496$. After 5, the next prime is 7, and $M_7 = 127$ is prime. This leads us to $64*127 = 8128$.

Armed with Euclid's Theorem, it might seem that we could go on generating perfect numbers forever. The trouble starts with the very next prime, 11: $M_{11} = 2047 = 23*89$ is not prime, so Euclid's Theorem does not apply. Euclid proved that *if M_p is prime, then $2^{p-1} * M_p$ is perfect*. Two thousand years later, Leonhard Euler (1707–1783) proved that *every even perfect number is given by Euclid's Theorem*; that is, every even perfect number corresponds to a prime p for which M_p is prime.

The perfect number question has evolved into one about prime numbers. For which primes p is M_p prime? These numbers have come to be known, not after Euclid or Euler, but after Marin Mersenne (1588–1648). In 1644, Mersenne announced that among the primes p between 31 and 257, M_p is prime precisely for

$$p = 31, 67, 127, \text{ and } 257$$

One hundred years later, Euler confirmed that M_{31} is prime. It took another century before it was confirmed that M_{127} is prime! Using new techniques that he developed himself, the French mathematician Edouard Lucas (1842–1891), proved that M_{127} is prime. In 1948, Oystein Ore, in his book, *Number Theory and Its History*, took the trouble to write out M_{127} in full:

$$M_{127} = 170141183460469231731687303715884105727$$

Ore explained that "the only reason for writing explicitly this huge number of 39 digits is that it is the largest number that has actually been verified to be prime." Don't misunderstand. There are infinitely many primes. There are primes that would make this one look like a drop in the bucket. In 1948, however, no one knew of a *specific* prime larger than M_{127}. But, the situation was about to change dramatically with the introduction of modern computers. Still, M_{127} is the largest number whose primality was determined without the aid of a computer. It is doubtful that this record will ever be broken.

In 1883, Pervouchine announced that M_{61} is prime. This startled the mathematical world because 61 is not on Mersenne's list! (In 1886, Seelhoff confirmed that M_{61} is prime.) Then, in 1903, an American mathematician, Frank Cole, showed that

$$M_{67} = (193707721) * (761838257287)$$

isn't prime after all! Mersenne was wrong again. In the period 1911–1952, it was shown (using the new method of Lucas) that among the primes p between 31 and 127, M_p is prime precisely for

$$p = 31, 61, 89, 107, \text{ and } 127$$

So, Mersenne was right about 31 and 127, but wrong about 61, 67, 89, and 107. (He was also wrong about M_{257}. It isn't prime either.) It seems ironic that these numbers should be named after the man who was the most spectacularly wrong about them.

As of this writing, the primes p between 128 and 100,000 for which M_p is prime are (precisely) p = 521, 607, 1279, 2203, 2281, 3217, 4253, 4423, 9689, 9941, 11213, 19937, 21701, 23209, 44497, and 86243. Moreover, M_{132049} is known to be prime but it is not known whether 132049 is the next prime p after 100,000 for which M_p is prime.

The Euclid/Euler Theorem characterizes the even perfect numbers. It is not known, however, if there are infinitely many of them, that is, if there are infinitely many Mersenne primes. It is also not known if there are any odd perfect numbers. However, Bryant Tuckerman has shown that if an odd perfect number exists, it must be greater than 10^{36}.

Exercises (6.6)

1. Find the prime factors of
   a. M_{13} b. M_{17} c. M_{19} d. M_{23}
   (*Hint:* Section 6.3, Exercise 6.)

2. Compute the 5th perfect number. Explain why the ancients did not find this number by trial and error.

3. Let $F(n) = 2^{2^n} + 1$. If $F(n)$ is prime, it is called a *Fermat prime*. Determine the primality of
   a. $F(0)$ b. $F(1)$ c. $F(2)$ d. $F(3)$ e. $F(4)$

4. In *Disquisitiones Arithmeticae*, published in 1801, C. F. Gauss proved that a regular polygon with n sides can be constructed using only a straightedge and compass, precisely when n is of the form

$$n = 2^k pq \ldots r,$$

   where $p, q, \ldots, r$ are different Fermat primes. (Here k may be 0, or n may be a power of 2.) Which values of n, $3 \leqslant n \leqslant 20$ are of this form? (See Exercise 3.)

5. A positive integer n is *almost perfect* if the sum of its divisors, other than n itself, is $n + 1$ or $n - 1$. Prove that every positive integral power of 2 is almost perfect. (*Hint:* Section 6.4, Exercise 14.)

6. Prove that n is perfect if and only if the sum of the reciprocals of all of its divisors (including n) is 2.

7. Show that the binary (base 2) representation of the perfect number corresponding to M_p consists of p 1's followed by $(p-1)$ 0's.

8. Suppose we start with a positive integer n. Add its divisors (except for n itself), obtaining a new number, m. Then add the divisors of m (except for m itself) to obtain another number. Continue the process in hopes of eventually getting back to the original number n. If $m = n$, then n is perfect. If the next number after m is n—that is, if the sequence ends after two steps—then m and n are said to be an amicable pair. Show that 220 and 284 are an amicable

pair. (During the Middle Ages these numbers were thought to promote friendship and love. *Hint*: Exercise 3, p. 209.)

9. The n-th triangular number is defined to be $1 + 2 + 3 + \cdots + n$. Which of the first 4 perfect numbers are triangular numbers?

Let n be a positive integer. Denote by s the sum of its positive integer divisors except for n itself. If $s<n$, then n is *deficient*. If $s>n$, then n is *abundant*.

10. By hand, without using the computer, determine which of the first 20 positive integers are deficient and which are abundant.

11. Write a computer program to determine which of the integers from 21 to 100 are abundant and which are deficient.

12. RUN the program you wrote for Exercise 11, and record the results.

13. Are there any odd abundant numbers? Write a program to search for one among the numbers less than 1000.

14. Twice a perfect number is an abundant number. Confirm this for the first three perfect numbers.

15. Half an even perfect number is a deficient number. Confirm this for the first three perfect numbers.

16. Show that there are infinitely many deficient numbers.

6.7 RATIONAL NUMBERS

A rational number is a fraction—a ratio, m/n, of integers in which $n \neq 0$. In this section, we are going to be interested in the decimal expansion of m/n when both m and n are positive. In particular, we will see that the decimal expansion of a number repeats if and only if the number is rational.

Suppose you were to divide m by n using long division. It might happen that the division process "terminates" at some point, leaving a remainder of 0. If we think, say, of 7.254 as 7.25400..., then, after awhile, the digits repeat. Namely, 0 repeats forever. In particular a "terminating" decimal expansion is a special case of a "repeating" decimal expansion.

What happens if the division doesn't terminate, if we continually produce nonzero remainders? After the decimal point, the digits of m have been passed by, and we are "bringing down" zeros in order to continue the process. Start taking careful note of the remainders. Since each of them is a number between 1 and $(n - 1)$, it is only a matter of time before one of them, say r, reoccurs. When you bring down 0 and divide by n, how will the result differ from the last time r was the remainder? It won't differ at all. You will begin to cycle through the same sequence of remainders, and record the same sequence of quotients, over and over and over again. That is, after a point, the digits start repeating. It may not be a single digit that repeats, but some sequence of (at most $n - 1$) digits will repeat and repeat indefinitely. Take, for example, $22/7 = 3.1427142714. \ldots$ The repeating sequence is 1427.

What about the other way around? Suppose you have a decimal expansion like .767676. . . in which 76 repeats. Check to see that this is the decimal expansion of 76/99. Similarly, .123123123. . . is the decimal expansion of 123/999.

What if the repeating part does not occur right away but shows up "after awhile"? Take, for example,

$$x = 1.23454545...$$

with 45 repeating. Then

$$100x = 123.454545...$$

$$= 123 + (45/99)$$

Divide both sides by 100 and add the resulting fractions. Then we see that

$$x = 123/100 + 45/9900$$

$$= 12222/9900,$$

a rational number.

Can we use the computer to produce decimal expansions? If you think about it for a minute, you will see that there is a technical problem. Even a number like

$$1/23 = .0434782608695652173913...,$$

has a repeating part 22 digits long. An instruction like

```
PRINT 1/23
```

isn't going to get us very far. The computer just doesn't carry enough digits for this purpose. We have to find a way to work around that limitation.

As a start, can we tell in advance how long it will be before the digits start repeating? Certainly we can give an upper bound. Once the digits of m have been passed by, we might have as many as $n - 1$ digits before a repeat. If there are L digits in m, then $2(n - 1) + L$ decimal places should always be adequate to display at least two complete segments of the repeating part (and frequently a lot more).

One way to count the number of digits in the integer m is to evaluate the LENgth of STR(m). Anticipating that we might want to cycle through the program several times, let's start off with

```
10  PRINT "THE DECIMAL EXPANSION OF M/N."
20  CLR
30  PRINT "PLEASE ENTER M,N."
40  INPUT M,N
50  K=2 * (N—1)+LEN(STR$(M))
```

Now, we might as well compute and PRINT the integer part of the quotient.

```
60  Q=INT(M/N)
70  PRINT Q;
```

The remainder will be $(M - N*Q)$, *before* we bring down a 0. After bringing down a zero it will be

```
80  R=10*(—N*Q)
```

Now, we want a FOR ... NEXT loop to compute some more of the digits. We could just PRINT each digit as it's computed, but then the computer will autmatically PRINT a space between each digit, something we can do without. One way to avoid the unwanted space is to turn the numeric character into a string character. This can be done in two ways. If Q is the current quotient, we can ask for STR(Q) or CHR$(48 + Q)$. Depending on your machine, these may produce identical responses, or one may still result in an unwanted space.

```
 90  FOR I=1 TO K
100  Q=INT(R/N)
110  D$=D$+CHR$(48+Q)
120  R=10*(R—Q*N)
130  IF R=0 THEN 150
140  NEXT I
```

The idea in line 110 is to produce a (long) string of digits to be PRINTed out later. This raises two concerns. One involves the "default value" of D\$ the first time through. The default value of a string variable which hasn't previously been used should be the "empty" string. But let's not take chances. Besides, we have already PRINTed the integer part of M/N. Let's go back and start the string, D\$, off as the decimal point.

```
85  D$="."
```

The second concern is this: How long can a string be? We are led to a (brief) discussion of "bits" and "bytes." A BIT is the smallest amount of information a computer can store. You might think of it as a switch that is either "on" or "off." When a BIT is on, it has the value 1. When it is off, it has the value 0. A BYTE is a group of (typically 8) BITS treated as a single unit. Thus, we can think of a BYTE as a string of eight 0's and 1's, a binary numeral.[†] Each alphanumeric character is represented in the machine by an 8-BIT code; it is stored as a single BYTE.

The string name (in our case D\$) may take up as many as 3 BYTEs. To this we must add 1 BYTE per character for the string. Counting on a total of no more than 255 BYTES per string, we should not expect this program to work on a denominator much larger than 100.

Resuming the program, we're ready to

```
150  PRINT D$+"..."
```

and wind things up

```
160  INPUT "ANOTHER RATIONAL";A$
170  IF LEFT$(A$,1)<>"N" THEN 20
180  END
```

One reason for using the negative formulation in line 170 is that simply pressing the ENTER key starts us through again. (The empty string <>"N".) If it's likely that you'll want to circle through the program several times, it is handier to do it this way.

[†] A BYTE is the base 2 representation of a number from 0 to 255. Base 2 arithmetic is as natural to a computer as counting on his/her fingers is to a 4-year old.

Exercises   (6.7)

1. Compute the decimal expansion of
   a. 5/9   b. 6/9   c. 7/9   d. 8/9

2. By hand, without using the computer, find the decimal expansion
   of $1/n$ for $n = 1, 2, \ldots, 10$. In each case, describe the repeating part
   of the expansion. In which cases is the repeating part of length
   $n - 1$?

3. Express the following as a fraction; that is, write as m/n where m
   and n are integers
   a. 0.111111...   b. 0.222222...   c. 0.444444...
   d. 0.232323...   e. 0.858585...   f. 0.769769...

4. Express as a fraction.
   a. 1.233333...   b. 3.146666...   c. 2.05202020...
   d. 6.4567567567...   e. 1.01001000100001...

5. Determine the repeating part of the decimal expansion of $1/n$ for
   $n = 11, 12, \ldots, 35$. (*Hint:* Use the program in the text.)

6. How might the program in the text be modified to accommodate
   denominators as large as 200?

7. When used to store a coded character, a BYTE of memory is commonly divided into two 4-BIT "zones." Each zone can be thought
   of as a 4-digit binary numeral, or as a single-digit hexadecimal
   (or HEX) numeral. (See Section 5.3, Exercise 12.) The EBCDIC†
   HEX code for the character W is E6. Since E(base 16) = 1110(base
   2) and 6(base 16) = 0110(base 2), the corresponding BYTE is
   [11100110]. Find the BYTES corresponding to the following EBCDIC
   HEX numerals.

   a. C8 (code for H)   b. D4 (code for M)
   c. E8 (code for Y)   d. 5B (code for $)
   e. 6F (code for ?)   f. 7C (code for @)

8. In the EBCDIC HEX code (See Exercise 7.), each alphabetic character, A–Z, corresponds to a two-digit numeral of the following
   type: A letter (C, D, or E), followed by a (base 10) numeral (1, 2,
   ..., 9). These HEX codes are arranged in increasing (base 16)

---

† Extended Binary-Coded Decimal Interchange Code.

numerical order, C1 through E9, except that E1 is missing. Thus, the EBCDIC HEX codes for A–Z consist of C1, C2, . . . , C9, D1, . . . , D8, D9, E2, . . . , E9, with E1 missing. (CA, CB, . . . , are also "missing.") Find the EBCDIC HEX code for:

a. B   b. E   c. I   d. O   e. U

9. Find the alphabetic characters corresponding to EBCDIC BYTES
   a. [11000110]   b. [11011001]   c. [11100111]
   (*Hint:* Convert the binary numeral to a hexadecimal numeral, and use Exercise 8.)

10. In the ASCII-8 HEX code, the alphabetic characters A–Z correspond to two-digit (base 16) numerals. In contrast to the EBCDIC HEX code (see Exercise 8), the ASCII-8 HEX code is sequential (no missing numerals) starting with A1 corresponding to the letter A and ending with BA corresponding to the letter Z. Find the alphabetic characters represented by the ASCII-8 BYTES
    a. [10101010]   b. [10110110]   c. [10101111]
    (*Hint:* Compare with Exercises 8 and 9.)

11. Write a program to have the computer identify and output the repeating part of the decimal expansion of m/n.

6.8 IRRATIONAL NUMBERS

One of the most revered names in mathematics is that of Pythagoras (ca. 582–500 B.C.).[†] Very little is known about the man except that he was a religious leader and philosopher as well as a great mathematician. One of his achievements was the discovery of a basic relationship between numbers and music.

[†] Pythagoras was roughly 25 years older than Confucius who lived a continent away. Pythagoras died just before the great war between the Greeks and Persians. Darius was defeated at the Battle of Marathon, one of history's pivotal events, in 490 B.C. The statesman Pericles and the playwright Sophocles were both born about 500 B.C. Pythagoras's portrait appeared on a coin, the silver tetradrachm, ca. 430 B.C. We will discuss his greatest triumph, the Pythagorean Theorem, in a later section.

A taut, vibrating string produces a musical tone or "note." Pythagoras found that the notes which sound harmonious with this ground note correspond to exact whole number divisions of the string. This discovery reinforced his predilection to view both the physical and spiritual worlds in terms of number. He became the leader of a numerological cult.

Great symbolism and purpose were attributed to positive integers by the Pythagoreans. The number 1 seems to have been associated with reason. Being the first number (after 1) that is a product of equals, 4 concerned justice. Perfect numbers were especially important to the Pythagoreans, and various roles were assigned to abundant numbers, deficient numbers, amicable pairs, primes, square numbers, and so on. This prejudice for whole numbers influenced mathematical thinking for the next 300 years.

The Greek mathematicians were inclined to think in geometrical terms. Starting with an arbitrary unit length representing 1, a segment twice as long was used to represent 2, a segment three times as long represented 3, and so on. Of course, it clearly is not true that every segment is an integer multiple of the unit segment. But, the early Greek mathematician/philosophers believed that, given any second segment, one could choose a *new* unit so that the old unit and the second segment are both integer multiples of the new unit. Maybe the old unit is 5 times the new one, and the second segment is 7 times the new unit. This is equivalent to saying that, in the original system of measurement (using the old unit), the length of the second segment is 7/5. In other words, the "doctrine of commensurable lengths" implies that any segment is a rational multiple of the unit segment. Stated purely in terms of numbers, the doctrine implies that every number is a rational number. This doctrine became an article of faith in the Pythagorean religious system. It is one of the small ironies of history that it should have been the Pythagoreans themselves, as mathematicians, who destroyed the foundation of their cult by the discovery of irrational numbers.

It is believed that the first number to have been proved irrational is $\sqrt{2}$. The proof is a classic example of the method known as "proof by contradiction." Suppose, to the contrary, that $\sqrt{2}$ were rational, say $\sqrt{2} = m/n$. Now, as you know, a rational number can be written in many different ways: 2/3, for example, is the same number as 4/6 or 6/9. We say that a fraction, m/n, is "reduced to lowest terms" if m and

n have no common factors, that is, if the GCD $(m,n) = 1$. In our proof we may assume, therefore, that $(m,n) = 1$.

The idea of the proof is to reach a logical contradiction. Specifically, we will show that attempting to express $\sqrt{2}$ as a fraction m/n, in which m and n have no common factors, leads to the conclusion that 2 is a common factor of m and n. This inconsistency proves that $\sqrt{2}$ is not a fraction. Proceeding with the proof, we first observe that $\sqrt{2} = m/n$ is the same as $\sqrt{2}*n = m$, and hence that

$$2n^2 = m^2$$

Now, the left-hand side of this equation is even, so $2 \mid m^2$. It follows from the Fundamental Theorem of Arithmetic that 2 must be a (prime) factor of m. Suppose $m = 2q$. Then

$$2n^2 = (2q)^2$$
$$= 4q^2$$

so

$$n^2 = 2q^2$$

Thus, n^2 is even. By the argument just used, it follows that 2 is a prime factor of n as well, that is, $2 \mid m$ and $2 \mid n$. Q.E.D.

This result presented the Pythagoreans with an acute dichotomy. Either they had to abandon their religious beliefs, or abandon the practice of mathematics. The two had become incompatible. Legend has it the cult split up. Since the mathematically inclined group was in the minority it took ship to a new location. The religious group is said to have maintained afterwards that their former colleagues all drowned when an offended diety sank the ship.

It is natural to wonder whether "irrational" numbers like $\sqrt{2}$ are common or rare. You will be asked in the exercises to prove that $\sqrt{3}$ is also irrational. In fact, if n is a positive integer, it can be shown that $\sqrt{n}$ is irrational unless n is a perfect square; that is, $\sqrt{n}$ is either a whole number or it is irrational. So, there are plenty of irrational numbers. Still, suppose you take all the numbers between 0 and 1, numbers like .254, .333. . . , $1/\sqrt{2}$, etc., and stir them up in a pot. Then pick one of the numbers from the pot at random. What is the probability that it will be a rational number?

In order to answer this question, we have to recall that the decimal expansion of a rational number must repeat "after awhile." (This makes it easy to write down additional irrational numbers. Just take care to write a decimal expansion that doesn't repeat. What about

$$x = 0.1011011101111011111\ldots?$$

Do the digits in x repeat after a time or not? Understand that numbers can have a pattern without having a repeating pattern.) With this observation at hand, we can answer our question. Let P be the probability that a randomly chosen number between 0 and 1 is rational. You will almost surely be surprised to learn that P = 0. Indeed, there are two surprises here. The first is that the fraction of the numbers which are rational could be 0. The second is that an event which is certainly possible can have zero probability, not just a very small probability, but 0 probability! These statements seem unbelievable.

Obviously, we can't mix the numbers in a pot and draw one out at random. But, we can describe an algorithm which, in principle, will produce the same result. Suppose we take 10 balls from a pool table to represent the digits 0 through 9. Put the balls into a paper bag, and draw one out at random. The number that appears on the ball is the first digit (after the decimal place) in our random number. Put the ball back in the bag, shake well, and draw again. The number appearing in the second draw becomes the second digit. Continue, in this way, to draw balls, record digits, replace balls, draw, record, replace, etc., being sure that each time a ball is drawn, it is equally likely to be any one of the 10. The question boils down to this: What is the probability that such a process will result in a repeating decimal? I don't mean one that repeats just once or twice, or just 100 billion times, but a sequence that repeats, repeats again, and again, and again, without deviating by so much as a single digit, for all eternity? (What is the probability that a random process is a repeating process?)

Perhaps this somewhat incredible turn of events will help us better appreciate the intellectual trauma that the discovery of irrational numbers inflicted on the Pythagoreans. Are you ready, psychologically, to deal with the implications of the argument just presented? Had you been a Pythagorean, would you have gone on board the ship?

Exercises (6.8)

1. Prove that between any two irrational numbers there is a rational number.

2. Let $S(n)$ denote the set of prime numbers p which (exactly) divide n. Show that $S(n) = S(n^2)$, i.e., show that the primes dividing n^2 are precisely the same primes that divide n. (*Hint:* Use the Fundamental Theorem of Arithmetic.)

3. Prove that $\sqrt{3}$ is irrational.

4. If x and y are irrational numbers, is their sum (always) irrational? Justify your answer.

5. If x and y are irrational numbers, is their product (always) irrational? Justify your answer.

6. If x is irrational is $1/x$ (always) irrational? Justify your answer.

7. Suppose x is rational and y is irrational. What can you say about
   a. $x + y$?   b. xy?   c. x/y?   d. y/x?

8. When we say $\sqrt{2}$ is about 1.414, we are approximating the irrational number, $\sqrt{2}$, with a rational number, $1.414 = 1414/1000$. Given any (terminating, decimal) approximation to $\sqrt{n}$, we can always obtain a better approximation. One way to do it was discovered by Isaac Newton. Here is Newton's algorithm.

   Step 1: INPUT n.
   Step 2: INPUT a positive approximation, x to $\sqrt{n}$.
   Step 3: Replace x with $x - [(x^2 - n)/2x]$.
   Step 4: PRINT x
   Step 5: If the new x is different from the old x, GO TO Step 3.
   Step 6: END.

   Write a program based on "Newton's Method" to obtain successively better approximations to $\sqrt{n}$.

9. RUN the program you wrote for Exercise 8, and record the successively better approximations to $\sqrt{2}$ when the initial value of x is:
   a. 1   b. 3   c. 10
   Compare the final approximation with SQR(2).

10. RUN the program you wrote for Exercise 8, and record the successively better approximations to $\sqrt{105}$ when the initial value of x is:
   a. 19   b. 11   c. 5   d. 1
   Compare the final approximation with SQR(105).

11. Newton's Method generalizes to the computation of the k-th root of n. The operative step becomes this one: Replace x with

$$x - \frac{x^k - n}{kx^{k-1}}$$

Write a revised program to INPUT n, k, and an initial approximation x, and to output successively better and better approximations to $\sqrt[k]{n}$.

12. RUN the program you wrote for Exercise 11, and record the output for:

a. $n = 64$, $k = 3$, $x = 1$ b. $n = 64$, $k = 6$, $x = 1$
c. $n = 2$, $k = 3$, $x = 1$ d. $n = 2$, $k = 5$, $x = 1$

Compare the best approximation you obtain with $n \uparrow (1/k)$.

Chapter 7
POLYNOMIALS AND FUNCTIONS

7.1 FUNCTIONS

Here is a code-breaking game you can ENTER and play. Each time you play the game, the computer generates a secret rule for encoding numbers. You play by ENTERing numbers. The computer will encode your number and output the result. After your second turn, the computer will ask you to guess the output. You win by guessing correctly twice in a row. But, you only get six turns. Moreover, ENTERing the same number twice will only waste a turn. (If your computer's RaNDom number generator requires a RANDOMIZE statement, you will have to add it to the program, say, in a new line 5.)

```
 10 REM BREAK THE CODE GAME
 40 M=-9+INT(19*RND(1))
 50 IF M=0 THEN 40
 60 B=1+INT(24*RND(1))
 70 DEF FN F(X)=M*X+B
 80 FOR I=1 TO 6
 90 INPUT "YOUR NUMBER";G(I)
100 IF I<3 THEN 290
110 FOR J=1 TO I-1
120 IF G(I)<>G(J) THEN 170
130 PRINT "YOU ALREADY TRIED";G(I)
135 PRINT "MY RESPONSE WAS";FN F(G(I))
140 PRINT "YOU WASTED A GUESS."
150 J=I-1
160 GO TO 300
170 NEXT J
180 PRINT "TRY TO PREDICT MY RESPONSE."
190 INPUT PG
200 IF PG<>FN F(G(I)) THEN 290
210 T(I)=1
220 IF T(I)+T(I-1)<2 THEN 270
230 PRINT "YOU GOT IT! MY FUNCTION WAS"
250 PRINT M;"X +";B
260 GO TO 320
270 PRINT "CORRECT! MY RESPONSE IS ";PG
280 GO TO 300
290 PRINT "MY RESPONSE IS ";FN F(G(I))
300 NEXT I
310 PRINT "BETTER LUCK NEXT TIME."
320 END
```

Play the game a few times until you're pretty good at it.

The new programming topic occurs in line 70. The BASIC statement DEF FN alerts the computer that you are about to DEFine a FuNction. The format for this statement is

```
DEF FN [name]([variable])=[expression]
```

In our case, the name of the function is F, while the variable is X. The "expression" is the "rule of assignment," which assigns to each number X the corresponding value F(X).

As you probably discovered, the fastest way to break the computer's secret encoding function is to ENTER first 0 and then 1. Since $F(X)$ = $M*X + B$, $B = F(0)$ is the "y-intercept" of (the graph of) $F(X)$, and $M = F(1) - B$ is its *slope*. In particular, you can always win the game in just four turns. To make the game more challenging, add these lines:

```
30  A=−3+INT(5*RND(1))
240 IF A<>0 THEN PRINT A;"X↑2 ";
245 IF A<>0 AND M>0 THEN PRINT "+";
```

and change line 70 to

```
70 DEF FN F(X)=A*X↑2+M*X+B
```

ENTER this program, but don't RUN it yet.

```
10 DEF FN UP(Z)=3*Z+4
20 FOR I=0 TO 5
30 PRINT "UP(";I;") =";FN UP(I)
40 NEXT I
50 END
```

Can you predict the output? RUN the program, and compare the actual response with what you predicted. (Note that the computer has to be reminded that UP is a function. Each use of UP must be preceded by FN. Without FN, the machine interprets UP(I) as a subscripted variable, that is, an entry from an array. Take FN out of line 30, and the computer will PRINT the default value, 0, for UP(I) on each pass through the loop. Try deleting FN from line 30 and see.)

Some versions of BASIC permit functions of more than one variable to be DEFined. Try RUNning this program to find out whether your version supports this additional capability.

```
10 DEF FN F(X,Y)=X*Y
20 PRINT FN F(1,2)
30 END
```

If you ENTERed the program correctly, and if an ERROR message occurs, you will know that you can't use functions of several variables. In many versions of BASIC, DEF FN cannot be used in direct mode. See how your machine responds to this sequence.

```
DEF FN F(X)=X↑2+3*X−5
PRINT FN F(2)
```

If it responds with an error message, you might be able to circumvent the difficulty by means of the following device. RUN

```
10 DEF FN F(X)=X↑2+3*X−5
20 STOP
```

Then try PRINTing FN F(2).

Finally, the DEF FN statement sometimes leads to curious ERROR messages. ENTER this program exactly as you see it here.

```
10 DEF FN F(X)=2X
20 PRINT FN F(2)
30 END
```

When you RUN the program, do you get an ERROR message for line 10 or line 20? The actual error is in the DEFinition of FN F(X). (What is the error?)

Exercises (7.1)

1. Find the slope and y-intercept of $f(x)$ =
   a. $3x + 2$ b. $2x + 3$ c. $-x + 4$ d. $6x$
   e. $-7x - 1$ f. 4   g. $x - .6$ h. 0

2. Find the value(s) of x for which $f(x)$ = 0 in each part of Exercise 1.

3. Without RUNning it on a computer, describe what the output from the following program would be.

```
10 DEF FN F(X)=3*X+2
20 FOR I=0 TO 1 STEP .2
30 PRINT I;F(I)
40 NEXT I
50 END
```

4. Without RUNning it on a computer, describe what the output from the following program would be.

```
10 DEF FN F(X)=X↑2−1
20 FOR I=−5 TO 5 STEP 1/2
30 PRINT I;FN F(I)
40 NEXT I
50 END
```

5. Write a program that will allow you to make use of the computer to prepare data points for graphing a function. (Make use of lines like

```
10 DEF FN F(X)=[EXPRESSION]
20 FOR I=A TO B STEP[S]
```

to be filled in later for specific functions.)

6. In each case, tabulate $(x, f(x))$, where $x = a$, $a+s$, $a+2s$, $a+3s$, ..., b.
   a. $f(x) = x^3 - 3x^2 - 9x + 6$; $a = -3$, $b = 5$, $s = 1$
   b. $f(x) = \ln(x)$, $a = .5$, $b = 10$, $s = .5$
   c. $f(x) = \text{EXP}(x)$, $a = -10$, $b = 10$, $s = 1$
   d. $f(x) = \sqrt{x}$, $a = 0$, $b = 2$, $s = .1$
   e. $f(x) = \sqrt{x}$, $a = 1$, $b = 10$, $s = 1$
   (*Hint:* See Exercise 5.)

7. Graph each of the functions in Exercise 6 on the corresponding interval $[a,b]$.

8. A "library function" that we haven't yet discussed is SIN. Tabulate $(x, \text{SIN}(x))$ where x goes from -6 to 6 in steps of $.4$.

9. Use your results from Exercise 8 to graph $y = \text{SIN}(x)$ on the interval $[-6,6]$.

10. Do you have a successful strategy for winning the more challenging version of the "Break the Code" game? Describe it.

7.2 POLYNOMIALS

A (real) *polynomial* is an expression of the form

$$p(x) = a_n x^n + a_{n-1}x^{n-1} + \cdots + a_1 x + a_0 , \qquad (7.1)$$

where n is a non-negative integer, and the a's are real numbers.[†] If $p(x) \neq 0$, then the highest power of x with a nonzero coefficient is called the *degree* of the polynomial. If $a_n \neq 0$ in Equation (7.1), then n is the degree of $p(x)$.

[†] The real numbers consist of the rational and irrational numbers combined.

When we substitute various real numbers for x in a polynomial, we are dealing with a *polynomial function*. A polynomial of degree 0 or 1 gives rise to a linear function. A polynomial of degree 2 is sometimes called a quadratic polynomial. The functions

$$f(x) = 3x^2 + 5x - 2$$
$$g(x) = 5.32x - \sqrt{2}x$$
$$h(x) = 1$$

are all polynomial functions, while

$$f(x) = \ln(x)$$
$$g(x) = 2^x$$
$$h(x) = x^{1/2}$$

are not.

The *y-intercept* of any function $f(x)$, polynomial or not, is $f(0)$. The x-intercepts, or zeros, on the other hand, are generally not so easy to find. In the case of a polynomial function, the zeros are called *roots*. Attempts to find general methods for obtaining roots of polynomials have occupied the attention of mathematicians for centuries.

For a polynomial function of degree 1, say $f(x) = mx + b$, the only root is $x = -b/m$. For a quadratic polynomial, we can "complete the square," or appeal to the *quadratic formula*. The zeros of $f(x) = ax^2 + bx + c$ are

$$x = [-b \pm \sqrt{b^2 - 4ac}]/2a$$

While there are similar formulas involving radicals for the roots of 3rd and 4th degree polynomials, there are no general formulas for degree 5 or higher. Not only does no one know of such a formula, no such formula can exist. This remarkable result follows from the work of Evariste Galois (1811–1832) and Niels Abel (1802–1829).

It turns out that the set of polynomials shares many properties with the set of integers. In particular, there is an analog of the Fundamental Theorem of Arithmetic. Every polynomial can be factored uniquely as a produce of "prime" polynomials. We are not going to study polynomials in detail, but we do need to know a few facts about them. One of the most useful of these is the following:

THEOREM:

Suppose $f(x)$ and $g(x)$ are polynomials. Then there exist unique polynomials $q(x)$ and $r(x)$ such that

$$g(x) = f(x)q(x) + r(x), \qquad (7.2)$$

where $r(x) = 0$ (that is, $f(x)$ is a factor of $g(x)$), or the degree of $r(x)$ is strictly less than the degree of $f(x)$.

This theorem asserts that, like integers, one polynomial can be divided by another to obtain a quotient (polynomial) $q(x)$ and a remainder (polynomial) $r(x)$.

The idea of the proof is to mimic the steps in long division. Suppose the degree of $g(x)$ is n and the degree of $f(x)$ is m. If $n<m$, then there is nothing to do; the quotient is 0, and the remainder is just $g(x)$. That is,

$$g(x) = f(x)*0 + g(x)$$

Otherwise, let $k_1 = n - m$, and consider the polynomial

$$g_1(x) = g(x) - f(x)*ax^{k_1},$$

where a is the ratio of the leading coefficient of $g(x)$ to that of $f(x)$. Then $g_1(x)$ is a polynomial of degree (at least) one less than the degree of $g(x)$. We have subtracted off the highest power term. (We may, inadvertently, have cancelled off some other powers as well.) Denote the degree of $g_1(x)$ by n_1. Rewriting the last equation, we obtain

$$g(x) = f(x)*ax^{k_1} + g_1(x)$$

Now, it may happen that $n_1<m$. In this case, $q(x) = ax^{k_1}$, $r(x) = g_1(x)$, and we are finished. Otherwise, let $k_2 = n_1 - m$ and consider

$$g_2(x) = g_1(x) - f(x)*bx^{k_2}$$
$$= g(x) - f(x)*(ax^{k_1} + bx^{k_2}),$$

or

$$g(x) = f(x)*(ax^{k_1} + bx^{k_2}) + g_2(x),$$

where b is the ratio of the leading coefficient of $g_1(x)$ to that of $f(x)$. Let $n_2 = $ the degree of $g_2(x)$. Since we have eliminated the highest power term in $g_2(x)$, $n_2 < n_1$. It may be that $n_2 < m = $ degree of $f(x)$. If so, we're finished. In this case, $r(x) = g_2(x)$, and $q(x) = $

$ax^{k_1} + bx^{k_2}$. Otherwise we continue. Since the n's keep getting (strictly) smaller at each stage, the process cannot go on forever. When it stops, we have our quotient and remainder.

Consider, for example, the case $g(x) = x^3 - 4x + 1$ and $f(x) = x - 2$. Then $n = 3$, $m = 1$, $k_1 = 3 - 1 = 2$, $a = 1$, and

$$g_1(x) = g(x) - f(x)*x^2$$

$$= (x^3 - 4x + 1) - (x - 2)*x^2$$

$$= 2x^2 - 4x + 1$$

Now, n_1 = degree of $g_1(x) = 2 > 1 = m$ = degree of $f(x)$. So, the process continues with $k_2 = 1$, $b = 2$, and

$$g_2(x) = g_1(x) - f(x)*2x$$

$$= (2x^2 - 4x + 1) - (x - 2)*2x$$

$$= 1$$

Now, n_2 = the degree of $g_2(x)$, is 0, which is less than 1 = the degree of $f(x)$. Hence, the process ends. Assembling the pieces gives

$$1 = g_2(x)$$

$$= g_1(x) - f(x)*2x$$

$$= g(x) - f(x)*x^2 - f(x)*2x$$

$$= g(x) - f(x)*(x^2 + 2x),$$

or

$$g(x) = f(x)*(x^2 + 2x) + 1,$$

that is, the quotient is $q(x) = x^2 + 2x$, and the remainder is $r(x) = 1$. Confirm it by multiplying out the right-hand side of

$$x^3 - 4x + 1 = (x - 2)*(x^2 + 2x) + 1$$

This example illustrates an interesting special case of the theorem, namely, when $f(x) = x - c$. In this case,

$$g(x) = (x - c)q(x) + r(x), \tag{7.3}$$

where $r(x) = 0$, or the degree, k, of $r(x)$ is strictly less than 1. Since k is a non-negative integer, it must be that $k = 0$. It follows that $r(x)$

$= r$ is just a number. What number? To find out, substitute $x = c$ in Equation (7.3). Then $g(c) = r$; that is,

$$g(x) = (x - c)q(x) + g(c), \tag{7.4}$$

for any polynomial $g(x)$ and number c. In particular, if c is a root of $g(x)$, then $(x - c)$ exactly divides $g(x)$. On the other hand, if $(x - c)$ were to exactly divide $g(x)$, then $g(x) = (x - c)q(x)$ for some polynomial $q(x)$. But then, certainly, $g(c) = 0$. Thus, we have proved the following.

COROLLARY:

Let $g(x)$ be a polynomial. Then c is a root of $g(x)$ if and only if $(x - c)$ is a factor.

To see one application of the corollary, consider the polynomial

$$g(x) = x^3 - 5x^2 + x + 3$$

We might happen to notice that the sum of the coefficients of $g(x)$ is zero, that is, $g(1) = 0$. If so, then the corollary tells us that $(x - 1)$ is a factor of $g(x)$. Knowing this, it is not hard to deduce the other factor:

$$g(x) = (x - 1)(x^2 - 4x - 3)$$

Now, the only way a product can be zero is if one of the factors is zero. The first factor produces the root $x = 1$. Any other roots of $g(x)$ must come from the second factor. From the quadratic formula, the roots of the second factor are:

$$[4 \pm \sqrt{16 + 12}]/2 = 2 \pm \sqrt{7}$$

Exercises (7.2)

1. Which is a polynomial?
   a. $2\sqrt{x}$ b. $\sqrt{2x}$ c. π^x d. x^π e. π f. x^2

2. Find the roots of:
   a. $x^2 - 2x + 1$ b. $x^2 - x - 20$ c. $x^2 + x - 1$
   d. $x^2 + 2x$ e. $3x + 6$ f. $2x^2 + 3x - 2$

3. Find the roots of:

   a. $3x^2 + 5x + 1$ b. $x^2 + x + 1$ c. $x^3 - 3x^2 + 2x$

   d. $x^3 - 2x^2 - 19x + 20$ (*Hint:* Find one root and factor.)

4. Find the roots of:

   a. $x^3 - 2x^2 - 5x + 6$ b. $x^3 - 2x + 1$

   c. $x^3 - 7x - 6$ d. $x^3 + 3x^2 + x - 2$

5. Write a program to find the roots of a quadratic polynomial.

6. Use the program you wrote in Exercise 5 to confirm your answer to

   a. Exercise 2a   b. Exercise 2b   c. Exercise 2c

   d. Exercise 2f   e. Exercise 3a   f. Exercise 3b

7. Write a program to input two polynomials and output their product.

8. Find the quotient and remainder when $g(x)$ is divided by $x - 1$, if $g(x)$ is:

   a. $x^3 - 3x^2 + 3x - 1$

   b. $x^3 - 1$

   c. $x^3 + 3x^2 + 3x + 1$

   d. $x^4 - 1$

   e. $x^4 + 1$

   f. $x^4 - 3x^3 + x^2 + x - 2$

9. Find the quotient and remainder when $x^4 - 3x^3 + x^2 + x - 2$ is divided by $x^3 + 3x^2 + 3x + 1$.

10. Use the program you wrote in Exercise 7 to confirm your answers to each part of Exercise 8. (*Hint:* Multiply your quotient by $(x-1)$. Then, by hand, add the remainder. See if you get the original polynomial back.)

11. Use the program you wrote in Exercise 7 to confirm your answer to Exercise 9.

7.3 RATIONAL ROOTS OF POLYNOMIALS

It was suggested in the last section that one way to find the roots of a polynomial, $p(x)$, of degree 3 is to guess one of them, call it c. Since

$p(x)$ can then be factored as a product of $(x - c)$ and a quadratic factor $q(x)$, the remaining roots of $p(x)$ can be obtained by using the quadratic formula on $q(x)$.

How easy is it to guess a root? In general, not very. But, there is a systematic method for finding a rational root (if there is one) in the special case that $p(x)$ has integer coefficients.[†]

THEOREM:

Let $p(x) = a_n x^n + \cdots + a_0$ be a polynomial of degree n with integer coefficients. If $c = m/k$ is a rational root of $p(x)$, with m and k relatively prime, then $m \mid a_0$ and $k \mid a_n$.

Since any rational number can be reduced to lowest terms, the assumption that $(m,k) = 1$ does not weaken the result.

Proof: If $p(m/k) = 0$, then

$$0 = a_n(m/k)^n + a_{n-1}(m/k)^{n-1} + \cdots + a_1(m/k) + a_0$$

Multiply both sides of this equation by k^n and rearrange the terms slightly to obtain

$$-a_0 k^n = a_n m^n + a_{n-1} m^{n-1} k + \cdots + a_1 m k^{n-1} \qquad (7.5)$$
$$= m(a_n m^{n-1} + a_{n-1} m^{n-2} k + \cdots + a_1 k^{n-1})$$

Since m (exactly) divides the right hand side, $m \mid a_0 k^n$. Since m and k are relatively prime, each prime divisor of m divides a_0. Moreover, if $p^r \mid m$ then $p^r \mid a_0$. Thus, $m \mid a_0$.

To prove that $k \mid a_n$, we write Equation (7.5) a little differently as

$$-a_n m^n = a_{n-1} m^{n-1} k + \cdots + a_1 m k^{n-1} + a_0 k^n$$
$$= k(a_{n-1} m^{n-1} + \cdots + a_1 m k^{n-2} + a_0 k^{n-1})$$

Since k divides the right-hand side, $k \mid a_n m^n$. Now, by the same argument used above, k must divide a_n. Q.E.D.

What do you suppose happens if the leading coefficient, a_n, is 1?

COROLLARY:

Let $p(x) = a_n x^n + \cdots + a_0$ be a polynomial of degree n with integer

---

[†] How likely is it that you would ever be able to guess an irrational root?

coefficients. If the leading coefficient, a_n, of $p(x)$ is 1, and if c is a rational root of $p(x)$, then c is an integer.

Proof. Write c as m/k, where m and k are relatively prime. Then, by the theorem, $k \mid 1$. Q.E.D.

A polynomial is said to be *monic* if its leading coefficient is 1. Suppose we're interested in the roots of the monic polynomial

$$p(x) = x^3 - 6x^2 - 8x + 7$$

If c is a rational root of $p(x)$, then c is an integer. There is more, namely, $c \mid 7$. Therefore, $c = 1, 7, -1$, or -7. Since the sum of the coefficients of $p(x)$ is -6; we see that 1 is not a root of $p(x)$. What about 7?

$$p(7) = 7^3 - 6*7^2 - 8*7 + 7$$

$$= 7(49 - 42 - 8 + 1).$$

So, $p(7) = 0$, and 7 is a root. Knowing this, we can easily factor $p(x)$:

$$p(x) = (x - 7)(x^2 + x - 1)$$

The remaining roots of $p(x)$ are the roots of $x^2 + x - 1$, namely, $[-1 \pm \sqrt{5}]/2$, that is, $.61803\ldots$ and $-1.61803\ldots$—numbers one is unlikely to guess off the top of the head.

Let's try another example, say,

$$f(x) = x^3 + 8x + 5$$

The only rational roots are integer divisors of 5. This limits us to 1, 5, -1, and -5. But wait. The coefficients of $f(x)$ are all positive. Thus, $f(x) > 0$, for any positive x. So, there is no point in even trying 1 and 5. Now, $f(-1) = -4$ and $f(-5) = -160$. It follows that $f(x)$ has no rational roots at all! Since there is no possibility of guessing an irrational root, we need to try another method or give up.

Consider this polynomial:

$$g(x) = x^3 - (3/2)x^2 - (1/2)x + 1/2$$

As it stands, we can apply none of the methods of this section. They all depend on the coefficients being integers, and the coefficients of $g(x)$ are not (all) integers. But, $g(c)$ is zero if and only if $2g(c)$ is zero. That is to say, the roots of $g(x)$ and of $2g(x)$ are identical. Thus, to find the roots of $g(x)$, we can look for the roots of

$$2g(x) = 2x^3 - 3x^2 - x + 1$$

This is not a monic polynomial so we can't use the corollary; but, we can use the theorem. If $c = m/k$ is a rational root, then $k \mid 2$ and $m \mid 1$. Thus, the possibilities for c are 1, 1/2, -1, and $-1/2$. Since the sum of the coefficients of $2g(x)$ is -1, $2g(1) \neq 0$. Let's try 1/2:

$$2g(1/2) = 2(1/2)^3 - 3(1/2)^2 - 1/2 + 1$$

$$= 1/4 - 3/4 - 1/2 + 1$$

$$= 0.$$

So, 1/2 is a root of $g(x)$. Thus, $(x - 1/2)$ is a factor. (If you like, you can equally well conclude that $(2x - 1)$ is a factor.) Once again, knowing what one factor is, we can easily obtain the other:

$$g(x) = (x - 1/2)(x^2 - x - 1).$$

Consider one final example:

$$h(x) = 28x^5 - 30x^2 - 45$$

Any rational root of $h(x)$ must be of the form m/k where $m \mid 45$ and $k \mid 28$; that is, we have to check 1, 3, 5, 9, 15, 45; 1/2, 3/2, 5/2, 9/2, 15/2, 45/2; 1/4, ..., 45/4; 1/7, ..., 45/7; 1/14, ..., etc., etc., and all the negatives of these! Clearly, it's time for a little help from the computer.

Exercises (7.3)

1. Suppose $p(x) = a_n x^n + \cdots + a_0$ is a polynomial with integer coefficients. Answer True or False. (True means always true. False means *not* always true.)
   a. $p(x)$ is monic if and only if $a_n = 0$.
   b. $p(x)$ has degree n if and only if $a_n \neq 0$.
   c. $p(x)$ cannot have an integer root if $a_n > 1$.
   d. If n is odd, then $p(x)$ must have a real root which might, however, be irrational.
   e. If all the coefficients of $p(x)$ are positive, then $p(x)$ has no positive roots.
   f. If all the coefficients of $p(x)$ are negative, then $p(x)$ has no negative roots.
   g. Since the coefficients of $p(x)$ are all integers, hence rational numbers, $\sqrt{2}$ cannot be a root of $p(x)$.

2. Justify your answers to Exercise 1. If you answered True, explain why. If you answered False, give a counter example.

3. Answer True or False.
   a. $x^2 + \sqrt{2}x - 3$ is a monic polynomial.
   b. The polynomial $x^3 + (3 + \sqrt{2})x^2 + (1 + 3\sqrt{2})x + 3$ can have no rational roots.
   c. If $f(x) = (x-c)q(x)$, where $q(x)$ is a polynomial, then, of course, $f(c) = 0$. But, there can exist (nontrivial) roots of a polynomial $f(x)$ that do not arise in this way.
   d. If $f(x)$ is any function, let $q(x) = f(x)/(x-c)$. Then $f(x) = (x-c)q(x)$. Hence, c is a zero of $f(x)$.
   e. If $f(x)$ is any polynomial, there exists a monic polynomial $g(x)$ such that $f(x)$ and $g(x)$ have exactly the same roots.
   f. If $f(x)$ is a polynomial with no rational roots, then $f(x)$ has no integer roots.

4. Justify your answers to Exercise 3. If you answered True, explain why. If you answered False, give a counter example.

5. By hand, without using the computer, find all rational roots of
   a. $3x^3 + 2x^2 + 2x - 1$
   b. $x^3 + 2x^2 - 12x + 15$
   c. $x^3 + x - 1$
   d. $x^3 - 3x - 2$
   (*Hint:* If you can find one root, c, then factor out $(x-c)$.)

6. Write a program to find all rational roots of
   $$28x^5 - 30x^2 - 45$$

7. Find all rational roots of:
   a. $10x^5 + 3x^4 + 29x^3 - 16x^2 + 17x - 3$
   b. $2x^4 + x^3 + 6x^2 - 2x + 3$
   c. $3x^5 + 8x^4 + 15x^3 + 50x^2 + 12x + 72$
   d. $36x^5 + 72x^4 - 13x^3 - 26x^2 + x + 2$
   e. $25x^6 - 126x^4 + 105x^2 - 4$
   (*Hint:* Modify the program you wrote for Exercise 6.)

8. Explain why $f(x) = 3x^8 - 5x^2 + 1$ must have an irrational root between 0 and 1.

9. Explain why $f(x) = 2^x$ has no rational zeros.

10. Using trial and error, see if you can find a rational zero of the computer's library function $f(x) = \text{SIN}(x)$. (*Hint:* See Section 7.1, Exercises 8 and 9.)

7.4 NEWTON'S METHOD

In Section 6.8, Exercises 8–12, we encountered a method for approximating the k-th root of a positive number, that is, for finding a root of the polynomial $x^k - c$, $c>0$. The method, due to Isaac Newton (1642–1727), can be used to approximate (real) roots of more general polynomials.

If $f(x)$ is the polynomial

$$f(x) = a_n x^n + a_{n-1} x^{n-1} + \cdots + a_1 x + a_0,$$

then the *derived* polynomial resulting from $f(x)$ is defined to be

$$f'(x) = n a_n x^{n-1} + (n-1) a_{n-1} x^{n-2} + \cdots + 2 a_2 x + a_1$$

Suppose you have a rough idea where a root of $f(x)$ might be, and suppose you start with a crude approximation, say x_1. Then Newton's Method for finding increasingly better approximations is this:

(1) INPUT x_1

(2) $x_2 = x_1 - f(x_1)/f'(x_1)$

(3) IF $x_2 = x_1$ THEN END

(4) PRINT x_1

(5) $x_1 = x_2$ (Replace x_1 with the current value of x_2.)

(6) GO TO STEP (2)

Let's implement Newton's Method on the computer for $f(x) = 2x^3 + 3x^2 - 12x + 6$. ENTER this

```
10  DEF  FN  F(X)=2*X↑3+3*X↑2-12*X+6
20  DEF  FN  DF(X)=6*X↑2+6*X-12
30  INPUT  "X1";X1
40  X2=X1-FN  F(X1)/FN  DF(X1)
50  IF  ABS(X2-X1)<10E-7  THEN  90
60  PRINT  X1
70  X1=X2
80  GO  TO  40
90  PRINT  "ROOT  IS  APPROX.  ";X2
100 END
```

Perhaps a bit of explanation would not be out of order regarding line 50. What we really want here is "IF X2 – X1 = 0 THEN 90". While we don't ever expect X2 to actually equal X1 (unless the root is a "terminating" rational), we do expect that, after some time, the increasingly better approximations might become indistinguishable to the computer. The discrepancy between the two approximations might be less than the number of digits carried by the machine. Unfortunately, due to round-off error, this time may never come. Surprisingly enough, computers often have difficulty recognizing 0 when they see it.

All that remains before RUNning the program is to choose X1. To get a very rough idea of what's going on, let's compute a couple of easy values for our polynomial, namely, $f(0)$ and $f(1)$. For the particular function of interest to us here, $f(0) = 6$ while $f(1) = -1$. This represents a fortunate state of affairs; the graph of $f(x)$ is above the x-axis at 0, and below it at 1. Evidently the curve crosses the x-axis somewhere between 0 and 1; that is, there is a root between 0 and 1. Since it seems closer to 1, let's begin with that. RUN the program and INPUT (X1 =) 1.

If you ENTERed the program correctly, you should now be looking at a DIVISION BY ZERO ERROR IN 40. The problem is that we have inadvertently INPUT an X1 for which DF(X1) = 0. We have stumbled on a root, not of $f(x)$ but of $f'(x)$. The algorithm clearly fails in this case. Before selecting another value to INPUT for X1, let's make a note of our "failure" with X1 = 1. It turns out, in most cases, that between two roots of a polynomial $p(x)$, there lies a root of $p'(x)$. If r is a root of $p'(x)$, and if X1<r, then Newton's method will typically "converge" to a root of $p(x)$ which is also less than r. If we choose X1>r, then the program will usually approximate a root of $p(x)$ greater than r. Let's try and see. RUN the program again, but this time INPUT 0.

It is expected that your computer PRINTed two or three preliminary approximations before settling on .653021. . . . This number, .653021 is the correct value, to six decimal places, of one of the roots of $f(x)$. RUN the program again, but this time INPUT (X1 =) 2. After three or four iterations, the computer will settle on another root of $f(x)$, namely, 1.322011. . . .

In principle, we could have obtained this value directly, without Newton's Method. Given that .653021 is (approximately) a root of $f(x)$, $(x - .653021)$ is, approximately, a factor. If we were to divide $f(x)$ by $(x - .653021)$, we would get a quotient, $q(x)$, and a (small) remainder, $r = f(.653021)$. The remaining roots of $f(x)$ are, approximately, the roots of $q(x)$.

What about the third root?† Since we have found two real roots, the third root must be real, too. (Non-real roots occur in complex conjugate pairs.) Consider this: If a, b, and c are the roots of a polynomial, $p(x)$, of degree 3, then

$$p(x) = d(x-a)(x-b)(x-c),$$

where d is the leading coefficient of $p(x)$. While it isn't fun, we can multiply the right hand side out, obtaining

$$p(x) = d[x^3 - (a+b+c)x^2 + (ab+ac+bc)x - abc]. \qquad (7.6)$$

The numbers

$$E_1(a,b,c) = a + b + c, \qquad (7.7)$$

$$E_2(a,b,c) = ab + ac + bc, \qquad (7.8)$$

$$E_3(a,b,c) = abc, \qquad (7.9)$$

are the so-called *elementary symmetric functions* of the roots of $p(x)$. In general, the r-th elementary symmetric function of n numbers is the sum of all nCr products of the numbers taken r at a time.

In our case, the leading coefficient, d, of $f(x)$ is 2. The next coefficient (of x^2) is 3. Thus, the missing root, call it c, must satisfy

---

† Perhaps the "third" root is equal to one of the ones we already have.

$$3 = -2(.653021 + 1.322011 + c)$$

$$-1.5 = 1.975032 + c$$

$$c = -3.475032$$

RUN the program a third time, INPUTting (X1 =) -4 and confirm that this is, indeed, the third root.

Does Newton's Method always work? Does it always work so fast? Will the method ever converge to a number that isn't a root? Why does it work at all? We've already seen the answer to the first of these questions. It certainly doesn't work if we inadvertently INPUT a root of the derived polynomial; and, of course, it won't produce a zero that isn't there. (A polynomial with real coefficients might not have any real roots.) The answer to the second question is "yes and no." One can contrive polynomials and initial guesses for which the method takes a long time. On the other hand, if there is a real root, and if you're trying to find it, Newton's Method works very well.

If Newton's Method converges, it converges to a root.[†] Finally, the best question is the last. In a sense, if you can answer it, you can answer all the others. Unfortunately, the answer to that question depends on techniques of differential calculus.

Exercises (7.4)

1. Find the derived polynomial, $f'(x)$, when $f(x) =$
   a. $x^4 + x^3 + x^2 + x + 1$
   b. $5x^3 - 2x^2 - 7x + 19$
   c. x
   d. $x^2 - 23$
   e. $x^{12} - 8$

2. Use the Newton's Method program to find as many roots as you can of the following:
   a. $x^3 - 3x^2 - x + 2$
   b. $x^3 - 3x^2 - 9x + 6$

[†] "Converges" means there is a number L such that the value of x get arbitrarily close to L if enough iterations are performed.

c. $x^3 + 3x^2 - 6x - 7$
d. $x^3 + 2x^2 - 3x - 1$
(*Hint:* You will need to make appropriate modifications to lines 10 and 20.)

3. Describe what happens when you implement Newton's Method for $f(x) = x^2 + 12$ starting with $x_1 = 2$. (What are the roots of $x^2 + 12$?)

4. Find as many roots as you can of the following:
   a. $x^3 + x^2 - x + 5$
   b. $x^3 + x^2 - x - 2$
   c. $x^3 - 3x^2 + 3x - 2$
   d. $28x^5 - 30x^2 - 45$

5. Find the first elementary symmetric function (E_1) of the roots of the following:
   a. $x^3 - 2x^2 + 3x - 4$
   b. $x^3 + x^2 - x + 5$
   c. $x^3 + x^2 - x - 2$
   d. $2x^3 - 6x^2 + x - 2$
   e. $3x^3 - x + 6$

6. Find the third elementary symmetric function (E_3) of the roots of each polynomial in Exercise 5.

7. Write a program to INPUT A, B, and C, and to output their first, second, and third elementary symmetric functions.

8. RUN the program you wrote in Exercise 7 and record the results if A, B, C =
   a. 1, 2, 3   b. 1, -1, 0   c. 1.52, .031, $-.86$

9. Given that 2 and -5 are roots of $x^3 - 10x^2 - 49x + 130$, use the
   a. first elementary symmetric function to find the third root.
   b. third elementary symmetric function to find the third root.

10. Repeat Exercise 9 for $x^3 - 10x^2 + 3x + 126$ given that 6 and -3 are roots.

11. Repeat Exercise 9 for $x^3 + 16x^2 - x - 16$ given that 1 and -1 are roots.

12. In the case of a monic polynomial, $p(x)$, of degree 4 with roots a, b, c, d,

$$p(x) = x^4 - E_1(a,b,c,d)x^3 + E_2(a,b,c,d)x^2$$

$$- E_3(a,b,c,d)x$$

$$+ E_4(a,b,c,d),$$

Find explicit formulas for $E_1(a,b,c,d)$, $E_2(a,b,c,d)$, ... (*Hint:* Multiply $(x-a)\,(x-b)\,(x-c)\,(x-d)$.)

13. Write a computer program to INPUT A, B, C, and D, and to output each of their (four) elementary symmetric functions. (See Exercise 12.)

14. RUN the program you wrote in Exercise 13, and record the results if A, B, C, D =
    a. 1, 2, 3, 4   b. -1, 0, 1, 2   c. 1, 1, 2, 3

15. Repeat Exercise 9 for $x^4 + 17x^3 - 39x^2 - 17x + 38$, given that 1, -1, and 2 are roots.

16. If x_1 is a root of $f(x)$, what value does Newton's Method produce for x_2.

Chapter 8
GEOMETRY

8.1 THE PYTHAGOREAN THEOREM

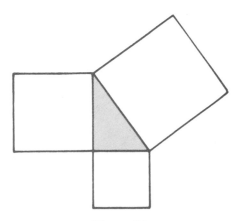

Figure 8.1

In 1877, the Italian astronomer, G. V. Schiaparelli[†] reported seeing markings, or lines, on the surface of Mars. He even produced rough drawings of these canali, or "channels." The leading Martian expert

---

[†]Schiaparelli (1835–1910) was Director of the Brera Observatory in Milan. His reputation was due, in part, to the discovery of the asteroid Hesperia in 1861. He also detected several double stars and presented evidence that Mercury and Venus rotate on their axes.

of the time, Percival Lowell,[†] contended that the sightings represented no less than canals, possibly bringing water from the Martian ice caps to thirsty cities. In arguing for a new 40-inch telescope at the University of Chicago, Samuel Leland wrote that it would be possible to see cities on Mars, "to detect navies . . . and the smoke of . . . towns."[‡] It wasn't *only* scientists whose imaginations were stimulated. Edgar Rice Burroughs entertained several generations with his Martian novels.

HELLO   2 + 2 = 4

Figure 8.2

At the turn of the 20th Century, nearly everyone was excited by the prospect of imminent communication with intelligent life on another planet. It was reasoned that, since we were trying to see the Martians through our telescopes, surely they must be looking at us through theirs. Proposals were made to send a conscious message. One idea was to plant trees in the barren steppes. Less patient folk advocated digging our own canals in the Sahara, filling them with oil, and touching it off one clear night. The only problem seemed to be the message. The ideas in Figure 8.2 are clearly unacceptable, as the Martians could hardly be expected to be acquainted with our numeration system or alphabet, much less our language. No, it would have to be something that the Martian skeptics could not explain away as accidental. And, it would have to be something that indicates clearly that we are intelligent ourselves, that we are sending a message *meant* to be read. Perhaps the best suggestion[§] is the one illustration in Figure 8.1 of the "Pythagorean Theorem."

PYTHAGOREAN THEOREM:

In triangle ABC, $a^2 + b^2 = c^2$, if and only if sides AC and BC are perpendicular to each other (see Figure 8.3).

[†]Lowell (1855–1916) established the Lowell Observatory in Flagstaff and predicted the existence of the planet Pluto. Among his numerous writings are *Mars and Its Canals* (1906), and *The Genesis of Planets* (1916).

[‡]See the February, 1973, issue of *National Geographic*.

[§]From the Teacher's Guide to Harold R. Jacobs's *Mathematics, A Human Endeavor*.

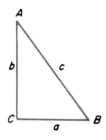

Figure 8.3

The angle formed by the intersection of two perpendicular lines is called a *right* angle. A *right triangle* is a triangle in which one of the angles is a right angle. The side of a right triangle opposite the right angle is sometimes called the *hypotenuse* of the right triangle.

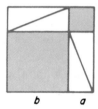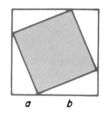

Figure 8.4

Of the more than 300 known proofs of the Pythagorean Theorem, the one illustrated in Figure 8.4 is probably close to the original proof given by Pythagoras in the 6th Century B.C. He is said to have offered 100 oxen to the Muses in thanks for the inspiration.

Special cases of the theorem were known much earlier. The builders of the great pyramids at Gizeh (2680 B.C.) knew that a builder's "square" could be produced by constructing a triangle with sides of lengths 3, 4, and 5. The ancient Babylonians had constructed hundreds of what we now call "Pythagorean Triples," namely, integer solutions to the equation

$$x^2 + y^2 = z^2$$

These are examples of right triangles in which the lengths of the three sides are all integers. Besides (3,4,5), other examples of Pythagorean Triples are (5,12,13), (8,15,17), and (20,21,29).

Asked to name the single most important theorem in all of mathematics, a plurality of mathematicians might well pick the Pythagorean Theorem. It establishes a profound yet elegantly simple relationship between geometry and number. Moreover, it affords many important applications. One of these is to the computation of distance.

Given a square of side 1, call the length of its diagonal h. Then

$$1^2 + 1^2 = h^2,$$

so

$$h = \sqrt{2}$$

In particular, it is precisely the great theorem that proves the existence of a number whose square is 2. Once this particular number was shown to be irrational, there could be no logical way to deny the existence of irrational numbers. (See Section 6.8.)

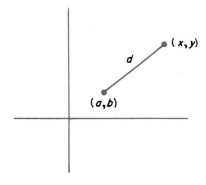

Figure 8.5

Consider the situation (illustrated in Figure 8.5) in which two points of the coordinate plane are given, say (a,b) and (x,y). Then, by the Pythagorean Theorem, the distance between them is

$$d = \sqrt{(x-a)^2 + (y-b)^2} \tag{8.1}$$

Exercises   (8.1)

1. Explain the significance of Figure 8.1. What would the Martians be able to deduce from it?

2. Explain how Figure 8.4 proves the Pythagorean Theorem.

3. Explain how Equation (8.1) follows from the Pythagorean Theorem and Figure 8.5.

4. Produce the hand calculations which confirm that the following are Pythagorean Triples:
   a. (3,4,5)   b. (6,8,10)   c. (9,12,15)   d. (9,40,41)

5. Find the length of the hypotenuse of a right triangle if the lengths of the sides are
   a. 2 and 3   b. 2.8 and 4.5   c. 5.5 and 4.8
   d. 2.99 and 1.80   e. 69.6 and 69.7   f. 33 and 56

6. Write a computer program to INPUT or READ three numbers and determine whether they constitute a Pythagorean Triple.

7. RUN the program you wrote for Exercise 5 and determine which of these is a Pythagorean Triple:
   a. (119,120,169)   b. (88,105,137)   c. (123,125,177)
   d. (123,126,175)   e. (220,459,509)   f. (3367,3456,4825)

8. Explain why it is obvious, without doing any calculation at all, that the triple of numbers in Exercise 7c couldn't be a Pythagorean Triple.

9. Write a computer program to INPUT or READ the coordinates of two points and to output the distance between the points.

10. Find the distance between
    a. (1,2) and (2,5)
    b. (1,8) and (5,3)
    c. $(-1,2)$ and $(5,-3)$
    d. $(-1,-4)$ and $(-3,5)$
    e. (1.2,3.4) and $(-.8, .6)$

11. Each of the following triples of numbers represent the lengths of the three sides of a triangle. Which of the triangles are right triangles?
    a. $\sqrt{3}$, 2, $\sqrt{5}$
    b. $\sqrt{2}$, $\sqrt{3}$, $\sqrt{5}$
    c. 9, 16, 25
    d. 396, 403, 565
    e. 1.28, 6.24, 6.58
    f. .85, 1.32, 1.57

12. Draw a careful, accurate picture of the triangle indicated in Exercise 11c.

13. Suppose (a,b,c) is a Pythagorean Triple. Explain why a and b cannot be equal.

14. Write a computer program to search for Pythagorean Triples involving positive integers $\leq n$.

15. What is $O(n)$ for your solution to Exercise 14? (*Hint:* See Section 6.3.)

16. Make a list of all Pythagorean Triples, (a,b,c), $1 \leq a < b < c \leq 30$.

17. Suppose that (a,b,c) is a Pythagorean Triple. Prove that (na,nb,nc) is a Pythagorean Triple, for any positive integer n.

18. Suppose that (a,b,c) is a Pythagorean Triple. If d is the greatest common divisor of a, b, and c, show that $(a/d,b/d,c/d)$ is a Pythagorean Triple.

19. A Pythagorean Triple (a,b,c) is said to be primitive if a, b, and c are relatively prime. Make a list of the primitive Pythagorean Triples with a, b, and c all less than 30.

8.2 ANGLES

Angles are commonly measured in two different units, degrees and radians. The degree measure is based on an (arbitrary) division of the circle into 360 parts.[†] Radian measure, on the other hand, is much more natural if less well understood. Choose your favorite unit of length (meters, feet, light-years). A unit circle is a circle whose radius is equal to the unit you have chosen. Consider an angle of interest to you and superimpose a unit circle on it so that the vertex of the angle is located at the center of the circle. (See Figure 8.6.) Then the measure of the angle is the length of the arc lying between the rays of the angle.

Radian measure has the advantage of making angle measure consistent with linear measure. Of course, instead of dividing the circle

---

[†] The fact that it is 360 and not some other number, say 100, surely dates back to the base 60 numeration system of the Babylonians . . . the same historical reason for dividing the hour into 60 minutes and the minute into 60 seconds. The ancients may have thought there to be just 360 days in a year.

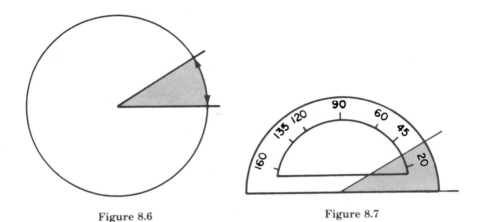

Figure 8.6   Figure 8.7

into 360 degrees, this approach divides it into 2π radians. In either case, angles are typically measured by means of a protractor such as the one pictured in Figure 8.7. (This one happens to be marked in degrees.) To convert from one system to the other, start with the basic relation, 2π radians = 360 degrees. Then,

$$1 \text{ radian} = 360/2\pi \text{ degrees}$$

$$\doteq 57.3 \text{ degrees}$$

and

$$1 \text{ degree} = 2\pi/360 \text{ radians}$$

$$\doteq .0175 \text{ radians,}$$

where "$\doteq$" means "about equal."

Two theorems from Euclidean Geometry are of interest to us here.

THEOREM 1:

The sum of the angle measures of any triangle is equal to two right angles. (In radian measure, the angle sum is π. In degrees, it's 180.)

THEOREM 2:

If two sides of a triangle have unequal length, then the longer side is opposite the larger angle.

An angle is *acute* if it is smaller than a right angle, if its radian measure is less than $\pi/2$. An angle larger than a right angle is *obtuse*.

And, a triangle containing an obtuse angle is said to be an *obtuse triangle*. It follows from Theorem 1 that a triangle can have at most one angle of measure $\geq \pi/2$. Thus, a triangle cannot be both right and obtuse.

Suppose you were given the coordinates of the vertices of a triangle. Let's write a program to determine if the triangle is a right triangle, an obtuse triangle, or neither.

```
10  FOR I=1 TO 3
20  PRINT "ENTER THE X,Y-COORDINATES OF"
25  PRINT "VERTEX";I
30  INPUT X(I),Y(I)
40  NEXT I
```

Now that we have the coordinates, let's compute the lengths of the sides:

```
50  A=SQR((X(2)−X(1))↑2+(Y(2)−Y(1))↑2)
60  B=SQR((X(3)−X(2))↑2+(Y(3)−Y(2))↑2)
70  C=SQR((X(3)−X(1))↑2+(Y(3)−Y(1))↑2)
```

Next, we need to know which is the longest side. Let's rename the sides, if necessary, so that C is the longest side:

```
80  IF A>C THEN Z=A:A=C:C=Z
90  IF B>C THEN Z=B:B=C:C=Z
```

Now, we know from the Pythagorean Theorem that $C^2 = A^2 + B^2$, if and only if our triangle is a right triangle. Indeed, $C^2 > A^2 + B^2$ if and only if our triangle is an obtuse triangle and $C^2 < A^2 + B^2$ precisely for triangles that are neither right nor obtuse. Here is a new programming gimmick:

```
100  K=1
110  IF C↑2>A↑2+B↑2 THEN K=2
120  IF C↑2<A↑2+B↑2 THEN K=3
130  ON K GO TO 140,160,180
140  PRINT "RIGHT TRIANGLE."
150  GO TO 200
160  PRINT "OBTUSE TRIANGLE."
170  GO TO 200
180  PRINT "ALL ANGLES ARE ACUTE."
200  END
```

The purpose of the variable K is to take the value 1 for right triangles, the value 2 for obtuse triangles, and the value 3, for the remaining triangles. The new BASIC code word is ON, used here with GO TO. Here is how ON works: If the value of K is 1, execution GOes TO the first line number listed. If the value of K is 2, execution transfers to the second line number listed, etc.

The ON statement can also be used with GOSUB. The general format for ON is

```
ON [variable] GO TO / GOSUB [line number],[line number], . . .
```

If the value of the variable is negative, the error message ILLEGAL QUANTITY will result. If the value of the variable is 0 or greater than the length of the list of line numbers, then the ON statement is ignored. If the variable is not an integer, the integer part is taken.

Let's test the program on some triangles that we know to be right triangles. (See Figure 8.8.)

1. Vertex coordinates are (0,0), (0,1), (1,0)
2. Vertex coordinates are (0,0), (1,0), (1,2)
3. Vertex coordinates are (1,0), (0,1), (1,1)

It may happen that the computer goofs. Whenever the machine does any arithmetic, there is the danger of round-off error. If the machine responded improperly to any of these examples, change lines 110 and 120 to

```
110  IF  C↑2—A↑2—B↑2>10↑—6  THEN  K=2
120  IF  A↑2+B↑2—C↑2>10↑—6  THEN  K=3
```

The problem now, of course, is that our identifications are only approximate; but, at least we're aware of it.

What about testing the program with an obtuse triangle, say

4. Vertex coordinates are (0,0), (1,0), (2,1)

or a triangle that is neither right nor obtuse:

5. Vertex coordinates are (0,0), (1,2), (2,0)

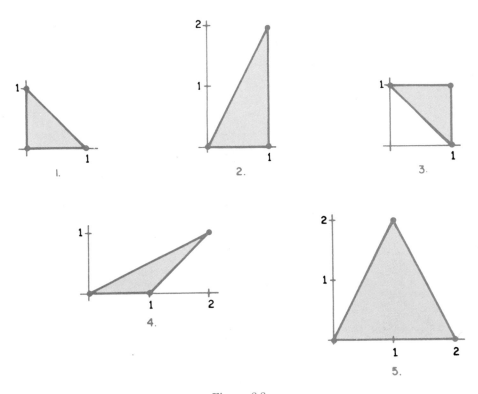

Figure 8.8

Here is a problem of a different sort, taken from *Ingenuity in Mathematics* by Ross Honsberger.[†] Suppose you are given two random numbers, say x and y, between 0 and 1. What is the probability that the three numbers, x, y, and 1, form the lengths of the sides of an obtuse triangle? (Note that, this time, we are starting with the sides, not the vertices.) The answer, astonishingly enough, turns out to be $(\pi - 2)/4$.

To begin, notice that we may interpret the pair (x,y) as the coordinates of a random point in the "unit square" of Figure 8.9. The probability that this point falls within any particular region, R, of the square is the ratio of the area of R to the area of the square. Since the area of the square is 1, the probability is just the area of R.

[†]Volume 23 of the New Mathematical Library published by the Mathematical Association of America, 1970.

In order for the three numbers to form a triangle at all, it is necessary (and sufficient) that $x + y > 1$. The region corresponding to $\{(x,y) : x + y > 1\}$ is pictured in Figure 8.10. Thus, the probability that the 3 numbers will lead to a triangle is $1/2$.

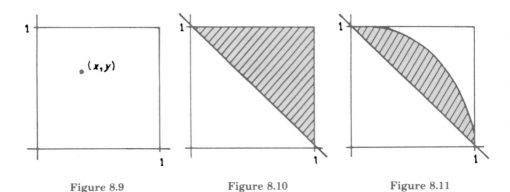

Figure 8.9   Figure 8.10   Figure 8.11

In order for the triangle to be obtuse, we need $x^2 + y^2 < 1^2 = 1$. The equation $x^2 + y^2 = 1$ is the equation of the unit circle centered at the origin. If $x^2 + y^2 < 1$, then (x,y) is *inside* the unit circle. Thus, for our three numbers to be the sides, not only of a triangle, but of an obtuse triangle, the point (x,y) must lie within the shaded region of Figure 8.11. Finally, the probability we want is just the area of this shaded region; but, that's easy to compute. The area of any circle is πr^2. In this case, $r = 1$. Our region comprises a fourth of the circle, minus the portion below the line $y = 1 - x$, (corresponding to the case in which no triangle is formed at all). Thus, the probability we seek is

$$\pi/4 - 1/2 = (\pi - 2)/4$$

Exercises (8.2)

1. Consider the program:

```
20 ON X*Y GO TO 70,40,60
30 PRINT "H";
40 PRINT "O";
50 PRINT "L";
60 PRINT "D"
70 END
```

What would the output be if you added one of the following lines?
a. 1 0 X = 1 : Y = 2   b. 1 0 X = 1 : Y = 1   c. 1 0 X = —1 : Y = 2
d. 1 0 X = 0 : Y = 1   e. 1 0 X = 2 : Y = 2   f. 2 5 X = 1 : Y = 2
(This is an exercise for you, not for the computer.)

2. About how many radians correspond to:
   a. 60 degrees   b. 45 degrees   c. 30 degrees   d. 18 degrees

3. About how many degrees correspond to:
   a. 0.6 radians   b. 0.1 radians   c. 5 radians
   d. 0.785 radians   e. 1.6 radians   f. 2.3 radians

4. Write a program to:
   a. convert degrees to radians
   b. convert radians to degrees

5. Write a program to INPUT the length of the hypotenuse and one
   (other) side of a right triangle and to output the length of the third
   side.

6. Draw a picture of the region in the unit square determined by the
   inequalities
   a. $x + y > 1$ and $x - y > 0$.
   b. $x - y < 0$ and $x^2 + y^2 > 1$.
   c. $x^2 + y^2 < 1$ and $x - y > 0$.
   d. $x + y < 1$ and $x - y < 0$

7. What is the probability that $X = RND(1)$, $Y = RND(1)$, and $Z = 1$
   will correspond to the lengths of the sides of
   a. a right triangle?
   b. a triangle that is neither right nor obtuse?

8. Write a program to determine empirically the probability that
   $X = RND(1)$, $Y = RND(1)$, and $Z = 1$ correspond to the lengths of the
   sides of an obtuse triangle.

9. RUN your program from Exercise 8, assuming a simulation of 1000
   trials. Compare the result with the theoretical prediction of $(\pi - 2)/4$.

10. Use your result(s) from Exercise 9 to approximate π.

11. A triangle with two equal sides is said to be *isosceles*. Write a
    program to INPUT the coordinates of the vertices of a triangle and
    to output whether or not it is isosceles.

8.3 SIMILAR TRIANGLES

Roughly speaking, two triangles having the same shape are said to be *similar*. In Figure 8.12, triangles *ABC*, *XYZ*, and *RPQ* are all similar to each other, but triangle *DEF* is similar to none of them. There are two (equivalent) ways to make the definition of similar triangles precise: In Geometry, two angles are *congruent* if they have the same measure.

(S1) Two triangles are similar if the angles of one can be matched up with the angles of the other so that corresponding angles are congruent.[†]

(S2) Two triangles are similar if the sides of one can be matched up with the sides of the other so that the lengths of corresponding sides are proportional.

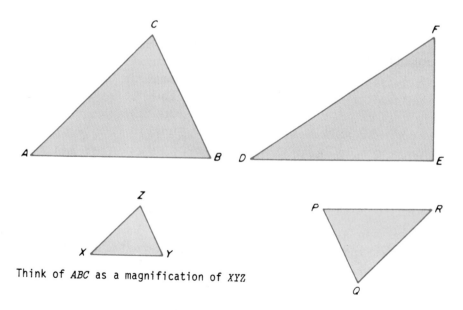

Think of *ABC* as a magnification of *XYZ*

Figure 8.12

[†] It doesn't matter what system of measurement is used as long as it is used consistently.

In Figure 8.12, angle A is congruent to angle X, angle B is congruent to angle Y, and angle Z is congruent to angle C. The second definition, on the other hand, seems a little ambiguous. The matching up is no problem. In Figure 8.12, side AB matches up with side XY, side BC matches up with side YZ, and side AC matches up with side XZ. Moreover (using XY, for example, to denote both a side of the triangle, and the length of that side),

$$AB/XY = BC/YZ = AC/XZ \qquad (8.2)$$

(The common value is the *power* of the magnification.) This is one interpretation of (S2). The other is, perhaps, less obvious:

$$AB/BC = XY/YZ, BC/AC = YZ/XZ, \text{ and } AB/BC = XY/YZ \quad (8.3)$$

But, (8.2) and (8.3) are equivalent as can easily be seen by cross-multiplication. For example,

$$AB/XY = BC/YZ \text{ and } AB/BC = XY/YZ$$

are two manifestations of the same identity.

Here is an application of similar triangles. Suppose you want to measure the height of a tree in your neighborhood without climbing to the top. Step outside on a sunny day with your yardstick and measure the length, S, of the shadow cast by the tree. Then measure the length, s, of the shadow cast by the yardstick. The height of the tree, measured in yards, is S/s. (See Figure 8.13. How are similar triangles used?)

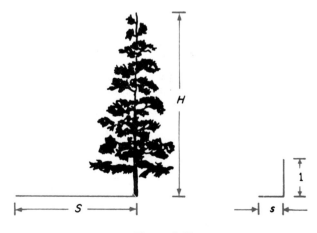

Figure 8.13

The equality of ratios indicated by (S2) is the basis for the subject of Trigonometry. We are going to trespass on that subject. Suppose you had two right triangles, each having an acute angle of measure a. Then the two triangles are similar! Matching up the right angles with each other and the two angles of measure a with each other leaves the third matched up. But, these third angles must both have the same measure, namely $\pi/2 - a$. So, they, too, are congruent. By (S1), the triangles are similar. See Figure 8.14 where, as is customary, the right angles are indicated by small squares.

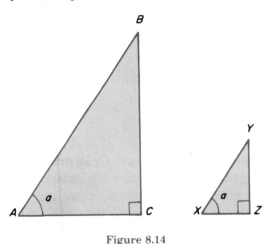

Figure 8.14

We can now make use of (S2). We are particularly interested in

$$BC/AB = YZ/XY \qquad (8.4)$$

Notice, first, that in both sides of Equation (8.4), it is the length of the hypotenuse which appears in the denominator. In the numerator we have, in each case, the length of the side opposite angle "a." What we have established is that the ratio is a function only of a. It does not depend on which right triangle we use. This ratio has a name. It is the *sine* of a, written $\sin(a)$.[†]

[†] The argument of these last two paragraphs was first given by Hipparchus (fl. 146–126 B.C.), the greatest astronomer of the ancient world. Hipparchus proved the equivalence of (S1) and (S2), and discovered the sine function, thus inventing trigonometry. Scientific Geography began with his invention of latitute and longitude. (In the northern hemisphere, latitude is the angle of elevation of the north star above the horizon.) Hipparchus is also credited with discovering the "precession of the equinoxes."

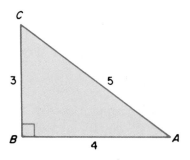

Figure 8.15

Curiously, it is often easier to determine the sine of an angle than to measure the angle itself. Consider the 3-4-5 right triangle pictured in Figure 8.15. The sine of angle A is $3/5 = .6$. The measure of angle A, on the other hand, turns out to be .6435 radians (or 38.8699 degrees).

If you have a protractor, it's no trouble at all to construct a right triangle containing any acute angle. Pick a number, a, between 0 and 1.57 ($\doteq\pi/2$).† Measure off an angle, O, of a radians, and drop a perpendicular from one of its sides to the other. Take a ruler and measure the length of the side opposite angle O. Divide by the length of the hypotenuse. That ratio is $\sin(a)$. It doesn't matter how big the triangle is, and it doesn't matter whether you measure lengths in feet, inches, meters, cubits, rods, or light years. The sine of a depends only on a. (In Figure 8.16, the sine of the angle at O is XA/XO. It is also YB/YO and ZC/ZO.)

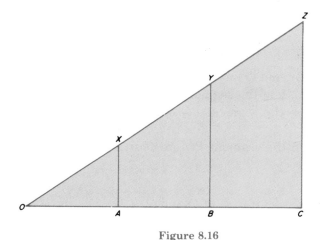

Figure 8.16

---

†Or, if you insist, pick a number between 0 and 90 (degrees).

It turns out that you can obtain sin(a) without going to so much trouble. One of the library functions on the computer is SIN.[†] It returns the sine of any angle measured in radians. Try ENTERing these:

```
PRINT SIN(1)
PRINT SIN(.7854)
PRINT SIN(.5236)
PRINT SIN(0)
```

Here, .7854 $\doteq$ $\pi/4$ radians and .5236 $\doteq$ $\pi/6$. (How many degrees do these correspond to?)

The common value of the ratios in Equation (8.2) can be thought of as a scaling factor. If the scaling factor happens to be 1, then the two triangles are said to be *congruent*; congruent triangles have the same shape *and* the same size. In particular, congruent triangles have the same area. (But, two triangles can have the same area without being congruent.)

Exercises *(8.3)*

1. Refer to Figure 8.12 in the text. Find XZ if:
   a. $AC = 3$, $BC = 2$, and $YZ = 1$
   b. $AC = 3$, $AB = 3.25$, and $XY = 1.25$
   c. $AC = 3.2$, $BC = 2.3$, and $YZ = 0.8$

2. Let $P = (a,b)$ and $Q = (x,y)$ be two points in the plane. Show that the coordinates of M, the midpoint of the line segment PQ are $((a+x)/2,(b+y)/2)$. (*Hint:* See Figure 8.5 and Figure 8.16. Use similar triangles.)

3. Let $P = (1,3)$, and $Q = (7,15)$. Find the coordinates of the
   a. midpoint, M, of the segment PQ.
   b. point T on the segment PQ one-third of the distance from P to Q.
   c. point F on the segment PQ one-fourth of the distance from P to Q.
   (*Hint:* Exercise 2.)

4. Repeat Exercise 3 if $P = (3,1)$ and $Q = (7,11)$.

---

† See Section 7.1, Exercises 8 and 9.

5. If triangle ABC is similar to triangle XYZ, and triangle XYZ is similar to triangle PQR, explain why triangle ABC is similar to triangle PQR.

6. Find the sine of the indicated angle in each part of Figure 8.17.

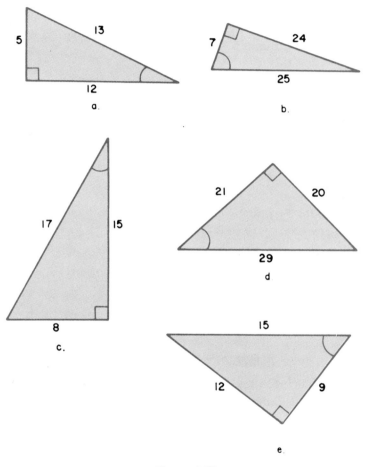

Figure 8.17

7. Let A be an acute angle. Explain why $0<\sin(A)<1$.

8. Consider triangle ABC in Figure 8.14. Prove that

$$[\sin(A)]^2 + [\sin(B)]^2 = 1.$$

(*Hint:* Use the Pythagorean Theorem.)

9. Write a program to confirm Exercise 8 by generating 10 random angles A (between 0 and $\pi/2$), computing B = $\pi/2$ − A, and outputting the left-hand side of the identity.

10. Consider a typical right triangle, ABC, where C is the right angle. (See, for example, Figure 8.14.) Then the *cosine* of angle A (denoted cos(A)) is defined to be the sine of angle B. Find the cosine of the indicated angle in each part of Figure 8.17.

11. If A is any acute angle, prove that

$$[\sin(A)]^2 + [\cos(A)]^2 = 1.$$

(See Exercise 10.)

12. One of the computer's library functions is COS, where COS(X) is the cosine of X, *provided* X is measured in radians. (See Exercise 10.) Find:
   a. COS(1)   b. COS(1.0472)   c. COS(0)   d. COS(1.57)

13. Write a program to confirm Exercise 11 by generating 10 random acute angles A and outputting the left-hand side of the identity. (*Hint:* See Exercise 12.)

14. Find COS(A), where A is
   a. 30 degrees   b. 60 degrees   c. 45 degrees
   (*Hint:* See Exercise 12. Convert to radians.)

15. It turns out that the sine function can be approximated by the polynomial

$$p(x) = x - x^3/3! + x^5/5! - x^7/7!,$$

provided x is measured in radians. Write a program to output the values of $\sin(x)$ and of $p(x)$ for the 15 values, $x = .1, .2, \ldots, 1.5$.

16. RUN the program you wrote for Exercise 15, and discuss the results. For which of the 15 values of x is the relative error, $|p(x) - \sin(x)|/\sin(x)$, the largest?

17. If you were to travel in an east-west direction, you would encounter lines of *longitude*. Longitudinal lines represent "great circles" around the Earth, passing through the poles. Starting at the "Prime Meridian" through Greenwich, England (longitude 0 degrees), and travelling west, the longitudinal lines increase to 180 degrees (at the International Date Line) and then decrease back to 0 at the Prime Meridian. (New York City lies at 74 degrees west longitude

and San Francisco at 122.5 degrees west longitude.) How far is it between two (whole number) degrees of longitude if:

a. your travel is along the equator? (The circumference of the Earth is about 24,890 miles.)

b. your travel is around a circle, centered at the North Pole, of radius 1 mile? (*Hint:* First compute the circumference of the circle, that is, the east-west distance "around the world" at this latitude.)

18. Using "spherical trigonometry," it can be shown that the east-west distance, in miles, between two degrees of longitude is 69*cos(latitude). Determine:

a. the distance between degrees of longitude, in an east-west direction, through your town. (*Hint:* First find out the appropriate latitude. See Exercises 10 and 17.)

b. the distance "around the world" at your latitude.

19. Consider Figure 8.18. Show that triangles *ABC*, *ACD*, and *CBD* are all similar.

20. Use Exercise 19 to provide an algebraic proof of the Pythagorean Theorem. (*Hint:* $AD/AC = AC/AB$.)

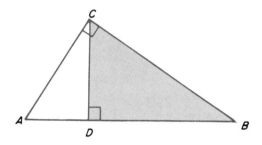

Figure 8.18

8.4 REGULAR POLYGONS

Two line segments having the same length are sometimes said to be congruent. An *equilateral triangle* is one in which all three sides are congruent.† Equilateral traingles comprise the first instance of what are known as regular polygons.‡

All three angles of an equilateral triangle are congruent. This follows automatically from the congruence of the sides: Not so with four-sided polygons. You can easily draw examples of quadrilaterals having four congruent sides but for which the four angles are not all right angles. In general, a *regular n-gon* is a polygon with *n* congruent sides and *n* congruent angles. A square is a regular 4-gon (or regular quadrilateral).

It is customary to use Greek prefixes to denote the number of sides of a polygon. For example, a pentagon has five sides, "penta" meaning five. Similarly, we can speak of a 6-gon or a hexagon. The familiar STOP sign is a regular octagon.

The first problem we want to tackle in this section is the (interior) angle problem. Each angle of a regular 3-gon (an equilateral triangle) measures $\pi/3$ radians. Each angle of a square measures $\pi/2$ radians. What about a regular pentagon, a regular hexagon, . . . ? The regular hexagon in Figure 8.19 has been divided into six (congruent) triangles. The sum of the measures of all six central angles is 2π, since the six together comprise one circle. Thus, each particular central angle must measure $2\pi/6$, or $\pi/3$ radians. This leaves $\pi - \pi/3 = 2\pi/3$ as the sum of the measures of the two remaining angles of any one triangle. But, by symmetry, these two angles are congruent. So, the measure of each of them is $\pi/3$.

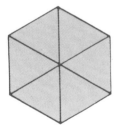

Figure 8.19

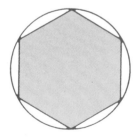

Figure 8.20

---

†If just two of the sides are congruent, the triangle is said to be *isosceles*. (The two angles opposite the congruent sides of an isosceles triangle are congruent as well.)

‡The word "polygon" literally means "many corners."

We can draw two conclusions from this discussion. The first is that each of the triangles in Figure 8.19 is equilateral. The second is that the measure of any (interior) angle of the hexagon (equal to the sum of the measures of the two noncentral angles of one of the triangles) is $2\pi/3$. Here is a summary of what we know about the angles of n-gons.

3-gon	4-gon	5-gon	6-gon	7-gon	. . .
$\pi/3$	$\pi/2$	?	$2\pi/3$	?	. . .

Any guesses for the angle of a pentagon?

We don't have to guess. We can use essentially the same argument that we just used for the hexagon. Take the corners of the pentagon as vertices of triangles, and its "center" as the common "central vertex" of all the triangles. Because the pentagon is regular, all five triangles are congruent. (This time, the triangles are isosceles but not equilateral.) Since each central angle comprises one fifth of the circle, it measures $2\pi/5$. This leaves

$$\pi - 2\pi/5 = 3\pi/5$$

for the *two* (congruent) non-central angles of each triangle. But, an (interior) angle of the pentagon consists of two of these angles. Thus, the measure of the angle of a pentagon is $3\pi/5$.

It is now clear how to generalize the method to any regular n-gon. First, carve it into n congruent isosceles triangles. The central angle of each triangle measures $2\pi/n$. Thus, the remaining two angles, together, measure

$$\pi - 2\pi/n = \pi(n-2)/n \qquad (8.5)$$

But, each angle of the polygon is comprised of two of these noncentral angles; that is, the measure of the (interior) angle of a regular n-gon is given by Equation (8.5).

Earlier, we referred to the "center" of a regular polygon. What we really mean by this is the center of the "circumscribed" circle. (See Figure 8.20.) Let's determine the perimeter of an "inscribed" regular polygon as a function of the radius of the circumscribing circle.

Consider an n-gon divided into n isosceles triangles, each with a central angle measuring $2\pi/n$. Take one of these triangles and find the midpoint of its "base." (The base is the side opposite the central angle.) Draw a line segment from this midpoint to the central vertex.

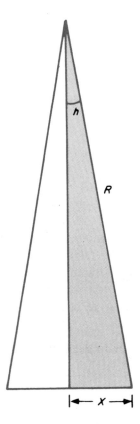

Figure 8.21

Then, by symmetry, our long, skinny triangle has been cut in half, forming two right triangles. (See Figure 8.21). The length of the hypotenuse of one of these right triangles is the radius, R, of the circle. The angle measure, h, is half a central angle, or π/n. Thus,

$$X/R = \sin(\pi/n)$$

or

$$X = R*\sin(\pi/n)$$

Now, $2X$ is the length of a side of the n-gon, so, the perimeter

$$P = 2nX \qquad (8.6)$$
$$= 2nR\sin(\pi/n).$$

In the case of the hexagon (Figures 8.19 and 8.20), we determined that the six triangles are equilateral triangles. Thus, the perimeter ought to be $6R$; that is,

$$6R = 2*6R\sin(\pi/6),$$

so

$$1/2 = \sin(\pi/6).$$

Confirm this by ENTERing

```
X=3.1415926535/6
PRINT SIN(X)
```

Now, $\pi/6$ radians = 30 degrees. It is worth remembering that this angle has sine exactly .5.

Suppose you were to inscribe a regular n-gon in a circle. If n were large, the difference between the circumference of the circle and the perimeter of the polygon would be small. In particular, the ratio of the perimeter of a regular n-gon to the diameter, $2R$, of the circumscribed circle should be approximately π; that is (see (8.6)),

$$\pi \doteq n*\sin(\pi/n) \tag{8.7}$$

Archimedes of Syracuse (ca. 287–212 B.C.) used this idea to estimate π. He calculated the perimeter of a regular 96-gon and divided by the diameter of the circumscribing circle. Using Equation (8.7), we can determine the value he obtained:

$$\pi \doteq 96\sin(\pi/96) \tag{8.8}$$
$$\doteq 3.1410$$

Naturally, Archimedes didn't obtain the answer the way we just did. We had to know what π was before we could compute the sine of $\pi/96$. Archimedes had to make *his* computation without knowing the value of π in advance. This explains why he chose 96 and not, say, 100. Carve a regular hexagon into six equilateral triangles as in Figure 8.19. Then cut each triangle in half (See Figure 8.21). Extend the central line until it meets the circle. If you do this for each of the six equilateral triangles, you will be able to construct a regular inscribed 12-gon. Cutting in half again yields a 24-gon, and twice more brings us to a 96-gon. Now, even without knowing what π was, Archimedes did know that $\sin(\pi/6) = 1/2$. Using this and a lot of calculation, he arrived at the answer in Equation (8.8).

Exercises (8.4)

1. Confirm that Equation (8.5) gives the measure of the (interior) angle of a regular n-gon when $n = 3, 4, 5,$ and 6.

2. What is the limiting value of Equation (8.5) when n grows arbitrarily large? What figure does the regular n-gon resemble more and more as n grows arbitrarily large?

3. Write a program to INPUT N and output the (interior) angle of a regular N-gon.

4. Suppose a regular N-gon is inscribed in a unit circle (that is, a circle of radius 1). Write a computer program to INPUT N and output the perimeter of the N-gon.

5. RUN your program from Exercise 4 and record the results if
   a. N = 100   b. N = 1000   c. N = 10,000
   Compare the outputs with 2π.

6. Consider a right triangle, ABC, with the right angle at C. Suppose the measure of angle A is $\pi/6$ radians = 30 degrees. If $BC = 1$, find AB and AC.

7. Consider a right triangle, ABC, with the right angle at C. Then the *tangent* of angle A (denoted $\tan(A)$) is BC/AC, that is, the length of the side opposite A divided by the length of the side (not the hypotenuse) adjacent to A. Compute the tangent of the indicated angle in each part of Figure 8.17, Section 8.3.

8. Show that $\tan(A) = \sin(A)/\cos(A)$. (See Exercise 7. See Section 8.3, Exercise 10 for the definition of $\cos(A)$.)

9. Show that $\tan(A) = \sin(A)/\sqrt{1 - \sin(A)^2}$. (See Exercise 8 and Section 8.3, Exercise 11.)

10. One of the computer's library functions is TAN: TAN(A) returns the value of the tangent of A, *provided* A is expressed in radians. (See Exercise 7 for the definition of "tangent.") Compute the following:
    a. TAN(.25)   b. TAN(.5)   c. TAN(.75)   d. TAN(1)

11. Write a computer program to confirm Exercise 8. Generate 10 RaNDom angles between 0 and 1.57, and compare TAN(A) with SIN(A)/COS(A).

12. Sometimes one knows the value of the trigonometric function, but not the measure of the angle itself. The computer library function ATN returns angles. If X is the tangent of an angle A, then ATN(X) is the measure of angle A in radians. Find the radian measure of

the indicated angle in each part of Figure 8.17, Section 8.3. (See Exercise 7.)

13. Find the degree measure of the indicated angle in each part of Figure 8.17, Section 8.3. (See Exercise 12. Convert from radian measure to degree measure.)

14. Write a program to INPUT Y, Y = sin(X), and to output X measured in degrees. (*Hint:* See Exercises 9, 12, and 13.)

15. Show that the area of the isosceles triangle pictured in Figure 8.21 is $X^2/\tan(h)$.

16. Show that the area of a regular n-gon, inscribed in a circle of radius R, is given by the formula

$$A = nR^2\sin(\pi/n)\cos(\pi/n).$$

(*Hint:* Exercises 8 and 15.)

17. Compute $n\sin(\pi/n)\cos(\pi/n)$ for
    a. $n = 100$ b. $n = 1000$ c. $n = 10,000$
    Explain why these numbers get closer to π as n increases.

18. The speed of light in air is about 186,000 miles per second. In water, however, light travels more slowly, causing light rays passing from one to the other to be "refracted." See Figure 8.22. The amount of refraction is given by Snell's Law: $\sin(a)/\sin(b) = v/w$, where v is the speed of light in air and w is the speed of light in water. If $a = 45$ degrees and $b = 32$ degrees, what is the speed of light in water?

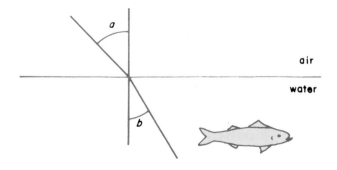

Figure 8.22

8.5 AREA

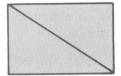

Figure 8.23

In a vague sense, the area of a plane region is the number of unit squares that can be packed into it. In particular, the area of a rectangle is height times width. In a formal, mathematical sense, we can essentially take this formula as the definition of area.

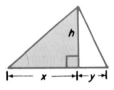

Figure 8.24

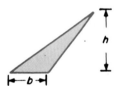

Figure 8.25

Given that we know how to calculate the area of a rectangle, we can calculate the area of any right triangle. Each of the triangles in Figure 8.23 comprises exactly half the rectangle. The area of a right triangle is half the base (that is, width) times the height. In Figure 8.24, the area of the right triangle on the left is $xh/2$, while the area of the one on the right is $yh/2$. Adding these two together† and factoring out the h yields $h(x+y)/2$, the height times half the base. It is not hard to obtain the same formula for the obtuse triangle in Figure 8.25.

Once we know how to calculate the area of any triangle, we can calculate the area of any polygonally shaped region such as the one

†In any formal, mathematical definition of area, we would state that area is non-negative, and that the area of the whole is equal to the sum of its parts. By parts, we mean subregions that overlap (at most) in line segments and points.

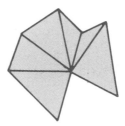

Figure 8.26   Figure 8.27

pictured in Figure 8.26. Simply carve the region into triangles, as in Figure 8.27, and compute the area of each triangle separately. One example of this procedure will both illustrate the method and provide us with an important and useful special case.

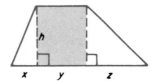

Figure 8.28

A *trapezoid* is a quadrilateral, two of whose sides are parallel. One kind of trapezoid is pictured in Figure 8.28. The area of this trapezoid is

$$A = xh/2 + yh + zh/2$$
$$= h(x/2 + y + z/2)$$

The second term in this last product is the average of the two parallel sides; that is,

$$[y + (x + y + z)]/2 = x/2 + y + z/2$$

The "formula" for the area of a trapezoid (any trapezoid) is the height times the average of the two parallel sides, where the height is the perpendicular distance between the parallel sides.

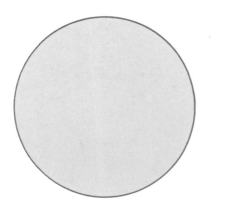

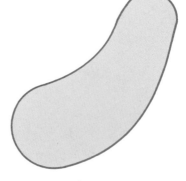

Figure 8.29   Figure 8.30

Now comes the hard part. What happens if the sides of the region are curved? In the case of the region depicted in Figure 8.29, the area is πr^2, where π is the well-known "fudge factor" $3.14159\ldots$; but, who knows the formula for the area of the region in Figure 8.30? The answer, of course, is that there is no formula. (No one has been brazen enough to suggest a formula involving a new fudge factor.) But, doesn't this cast a little doubt on the formula for the area of a circle? None of the other formulas involves mystery numbers. What is π, anyhow? The exact identity of π has eluded us time and time again. The great logician, Bertrand Russell (1872–1970), wrote in *A Mathematician's Nightmare*, "Pi's face was masked, and it was understood that none could behold it and live. But, piercing eyes looked out from the mask, inexorable, cold, and enigmatic." Russell also suggested the following mnemonic device:

> "Sir, I bear a rhyme excelling
>   In mystic force and magic spelling
>   . . . "

(Replace the comma with a decimal point, and count the number of letters in each word.)

What chance is there of carving the region in Figure 8.30 into subregions whose areas we can calculate? None. But, in many cases, we can obtain very good approximations by using the *Trapezoidal Rule*: Suppose we are interested in a region that is *almost* a rectangle, failing only in that the top portion is curved. The idea of the Trapezoidal Rule is to carve that region into smaller "almost rectangles" and to approximate the top of each of these by a straight line. (See Figure 8.31.)

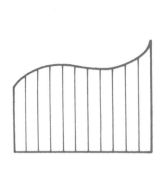

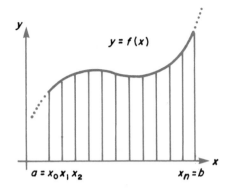

Figure 8.31   Figure 8.32

Suppose we are lucky enough to know a function, say $f(x)$, whose graph is the curvy top of the almost rectangle from $x = a$ to $x = b$. Then we can divide the segment, $[a,b]$, into n equal subintervals $[x_0,x_1]$, $[x_1,x_2]$, ... of width $w = (b-a)/n$. (Here, $a = x_0$, and $b = x_n$.) Finally, we can calculate the area of each trapezoidal approximation and sum them to obtain:

$$A = w*[(f(x_0)+f(x_1))/2] + w*[(f(x_1)+f(x_2))/2] + \cdots$$

$$= (w/2)*[f(x_0) + f(x_1) + f(x_1) + f(x_2) + f(x_2)$$

$$+ f(x_3) + \cdots]$$

$$= (w/2)*[f(x_1) + 2f(x_2) + 2f(x_3) + \cdots + f(x_n)]$$

This somewhat hideous (looking) expression is the *Trapezoidal Rule* for approximating areas of almost-rectangles. While a region such as the one picutured in Figure 8.30 cannot be cut into triangles, it can be divided into rectangles and almost-rectangles. Thus, the Trapezoidal Rule provides a general approach to finding approximate areas.

Suppose you are given the task of finding the area of a circle of radius 1 as a means of evaluating π. The unit circle, centered at the origin, has the equation $x^2 + y^2 = 1$. Thus, for the top half at least, we can describe the circle as the graph of the function

$$y = f(x) = \sqrt{1 - x^2}$$

By symmetry, it suffices to calculate the area of the region under this graph from $x = 0$ to $x = 1$ (and multiply the result by 4). Let's write a program to do this computation by means of the Trapezoidal Rule.

```
10  REM  TRAPEZOIDAL  RULE
20  A=0:REM  LEFT  HAND  ENDPOINT
30  B=1:REM  RIGHT  HAND  ENDPOINT
40  DEF  FN  F(X)=SQR(1-X↑2)
```

Now, we need to decide how many subintervals to use. Let's INPUT N.

```
50  INPUT  "NUMBER  OF  SUBINTERVALS";N
60  W=(B-A)/N
```

Next, we initialize a variable AR, which is going to evolve into the area of a quarter circle.

```
 70  AR=0
 80  FOR  I=1  TO  N-1
 90  AR=AR+2*FN  F(A+I*W)
100  NEXT  I
```

Now add the first and last terms, $f(x_0) = f(a)$ and $f(x_n) = f(b)$. Finally, multiply by $(b-a)/2 = $ W/2.

```
110  ARE=AR+1+0
120  AREA=ARE*W/2
130  PRINT  "AREA  =";AREA
140  END
```

Of course, the computer may recognize only the first two symbols in a variable name. To the computer, AR, ARE, and AREA may all represent the same variable. Perhaps it is a little frivolous to use ARE and AREA.

Since we are after π, the area of the whole circle, let's add the following

```
135  PRINT  "PI  =";4*AREA
```

RUN the program for $n = 100$ and for $n = 1000$. (Record the results and you will have done Exercises 1a and 1b.)

Exercises   (8.5)

1. RUN the program in the text for:
   a. $n = 100$ b. $n = 1000$ c. $n = 10,000$

2. Find the approximate area under the curve $y = 9 - x^2$ between $a = 0$ and $b = 2$. (*Hint:* See Figure 8.33. Modify the program in the text. Let $n = 1000$.)

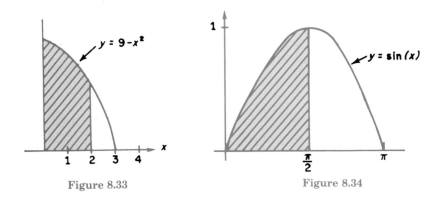

Figure 8.33   Figure 8.34

3. Find the approximate area under the curve $y = 9 - x^2$ between $a = 0$ and $b = 3$. (*Hint:* See Exercise 2. Let $n = 1000$.)

4. Find the approximate area under the curve $y = \sin(x)$ between $a = 0$ and $b = \pi/2$. (*Hint:* See Figure 8.34. Modify the program in the text. Let $n = 1000$.)

5. Use the Trapezoidal Rule to find the approximate area of the ellipse $10x^2 + 15y^2 = 150$. (*Hint:* See Figure 8.35.)

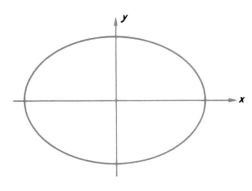

Figure 8.35

6. Compute the exact area of the ellipse in Exercise 5, and compare with the approximation given by the Trapezoidal Rule.

7. Heron's formula for the area of a triangle in terms of its 3 sides a, b, c, is

$$A = \sqrt{s(s-a)(s-b)(s-c)}$$

where $s = (a + b + c)/2$. Confirm Heron's formula for each of the right triangles in Section 8.3, Figure 8.17.

8. Write a program to INPUT the coordinates of the three vertices of a triangle and to output its area. (*Hint:* See Exercise 7.)

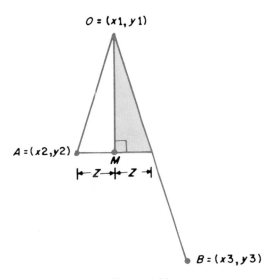

$O = (x1, y1)$

$A = (x2, y2)$

M

$\leftarrow Z \rightarrow|\leftarrow Z \rightarrow|$

$B = (x3, y3)$

Figure 8.36

9. Write a program to INPUT the coordinates of the vertex, O, and the coordinates of points A and B on each side of an arbitrary angle, and to output the degree measure of the angle. (*Hint:* See Figure 8.36. First compute Z. Then the radian measure of the angle is $2*\text{ATN}(Z/MO)$. See Section 8.4, Exercises 7, 9, 12, and 13.)

10. In Section 4.5, particularly Figure 4.5, we saw that approximately 34% of normally distributed data lies between the mean and one standard deviation above the mean. The normal curve is described by the function

$$f(x) = \frac{1}{\sqrt{2\pi}} e^{-x^2/2}$$

where the variable x is given in units of standard deviations. It turns out that the 34% figure comes from computing the area under the graph of $f(x)$ from $a = 0$ to $b = 1$. Use the Trapezoidal Rule to compute an approximate value for this area. (*Hint:* Modify the program in the text. Choose N = 1000.)

11. Confirm the other percentages given in Section 4.5, Figure 4.5. (See Exercise 10.)

12. Confirm the following entries from Table 4.1, Section 4.6: The fraction of normally distributed data between the mean and
    a. .4 standard deviations above the mean is 15.5%.
    b. 1.3 standard deviations above the mean is 40.3%.
    c. 2.6 standard deviations above the mean is 49.5%
    (See Exercise 10.)

INDEX